Robot Technology

Volume 5: Logic and Programming

Robot Technology

A Series in Eight Volumes

Series Editor and Consultant: Philippe Coiffet
English Language Series Consultant:
I. Aleksander, Brunel University, Uxbridge, England

Volume 5

LOGIC AND PROGRAMMING

Michel Parent and Claude Laurgeau

Kogan Page
London

Translated by Meg Tombs

First published 1983 by Hermes Publishing (France)
51 rue Rennequin, 75017 Paris, France

English language edition first published 1984
by Kogan Page Ltd, 120 Pentonville Road, London N1 9JN

Acknowledgements
The authors would like to thank the numerous colleagues from France and abroad for their valuable contributions. In particular, they are indebted to the robotics specialists at Peugeot-Citroen, M. Caillot, Girard and Quéromes.

Special thanks are due to Marc Bourez from Automatique Industrielle for his contribution to the writing of Chapter 5.

British Library Cataloguing in Publication Data
Parent, Michel
Robot technology.
Vol. 5 : Logic and programming
1. Robots, Industrial
I. Title II. Laurgeau, Claude
629.8'92 TS191

ISBN 0 85038 650 0

Printed and Bound in Great Britain by
T. J. Press (Padstow) Ltd, Padstow, Cornwall

Contents

Foreword

In contrast to the effortless ease with which human beings control their limbs, the design of controllers for robotic manipulator arms is a detailed, meticulous business. Motors controlling the arms need to be started and stopped at just the right moment so that the performance demanded by the user may be achieved at the end of a complicated manoeuvre. And yet, the same user wishes to express the task for the robot in the simplest possible terms without reference to the minute details of control sequences that his task demands. It is the design of such interfaces between man and machine that is the subject of this volume.

Parent and Laurgeau develop the subject in a direct and logical order. They first explain the principles of maximal effort control which not only ensure that motors are driven to provide high accuracy, but also that this should be done with the least waste of energy and in the shortest possible time. In this context, they describe the operation of pneumatic logical devices that make rapid decisions at power levels that exceed, by several orders, those that can be achieved with electronic devices. They achieve this whilst keeping the reader aware of the logical principles that are involved in the design of master control units: the devices responsible for appropriate actions being taken as a function of time.

But central to these master control systems is the provision of an appropriate programming language which gives the user the means whereby he can express the specification for the robot task. The authors survey the pros and cons of various available programming languages and styles, always within the framework of the logic of sequence control. They also pay attention to sensors and end effectors as these too must both provide information for the programs and receive the result of the logical decisions.

Appropriately, the book ends by stepping away from the system itself and paying attention to methods for its design. Computer-aided methods and simulation systems are discussed, providing the reader with guidelines as to how such schemes are selected and used. This volume is not only an important part of the *Robot Technology Series*, but also, in its own right, makes a valuable contribution to the understanding of designing robotic systems

Igor Aleksander
May 1984

Chapter 1

Introduction

Robots became real in industrial terms with the appearance of the first versatile and reprogrammable machines, which were able to do the work normally done by man in many industrial tasks. The versatility of these robots arose from the fact that they were not designed *a priori* for a particular task (although robots produced nowadays are becoming increasingly more specialized, a parallel also found in the human work-force). These robots can handle a range of diverse components and can also manipulate the tools required to carry out tasks characterized by complex trajectories. Versatility is further increased when movement of the robot joints can be changed rapidly to adapt to variations in work or to an entirely different task.

Like the numerically controlled machine tool from which it derived some of its operating principles, the Devol and Engelberger Unimate-type industrial robot took a number of years to become accepted in industry. Fifteen years after the development of the first prototype, Unimation, the world leader in the field and at the time virtually the only manufacturer in the market, had produced only about 1,000 machines.

Then, the Japanese, who had just formed the Japanese Industrial Robot Association (JIRA), announced in 1975 that they had 65,000 robots at work in industry. These figures, which were widely published in the Western press, promoted Japanese industrial techniques. The number of fact-finding visits organized by the West at the time to Japan is ample proof of this. Although there can be no doubt that industrial robotics was more rapidly adopted by Japanese industry than in the other industrialized countries, it was soon realized that the differences between Japanese and other industrial practices were not as great as was first believed. The Japanese definition of what was a 'robot' accounted for these very high figures, and the effect was to promote their industry and modernization techniques.

Western industrial countries quickly reacted to JIRA's figures with more precise definitions and evaluations of the world robot population. Among the most active robot associations similar to JIRA, one must mention the Robot Institute of America (RIA), the British Robot Association (BRA) and the French Industrial Robot Association (AFRI).

Founded in 1974, the Robot Institute of America (RIA) now has a membership of over 190 major companies.

1.1 Definitions

The definition of a robot is necessarily vague since it relates to human perception of machine anthropomorphism. Therefore, a universal definition is impossible to establish, although many attempts have been made. Here are some:

> 'Apparently human automaton, intelligent and obedient but impersonal machine' (*The Concise Oxford Dictionary*).

This definition does not accurately describe the majority of robots used in industry today.

> 'Versatile, flexible mechanism with displacement functions similar to those of human limbs or with displacement functions controlled by sensors and means of recognition' (JIRA, 1980).

This definition is vague and includes maximal effort manipulators, remote-controlled manipulators, prostheses etc.

> 'A robot is a reprogrammable multifunctional manipulator designed to move material, parts, tools, or specialized devices, through variable programmed motions for the performance of a variety of tasks' (RIA, 1979).

Here, the term *manipulator* is not defined. Does a single axis mechanism qualify as a manipulator? Does a computer plotter or a machine tool fall into this category?

> 'An industrial robot is a reprogrammable device designed to manipulate and transport parts, tools or specialized manufacturing implements through variable programmed motions for the performance of specific manufacturing tasks' (BRA).

The definition which is the most accurate is without doubt given by AFNOR (registered standard NF E 61-400 in August, 1983 and presented at international level ISO for acceptance). This specification is presented in detail in the appendices to this chapter. AFNOR is careful to define the term *manipulator* as a 'multifunctional mechanism with several degrees of freedom controlled by a human operator or by an automatic system' and *industrial robot* as an 'automatic manipulator that is under positional servocontrol, reprogrammable and versatile'.

Despite these efforts, the AFNOR definition is not sufficiently detailed to allow systematic classification of machines as being or not being robots. There will always be an element of doubt when attempting to define some machines as robots (eg numerically controlled machines).

In response to the need for a comprehensive classification covering the range of robots at work in France, the Association Française de Robotique Industrielle (AFRI) established a guide using examples to define each class of machine used in their automated industries. This empirical and arbitrary method is so far the only satisfactory means of providing a working definition.

1.2 The AFRI classification

AFRI has created three basic categories in its classiciation of industrial robotics: *manual handling devices, automatic handling devices* and *robots.* In all categories the devices must include a means of gripping or a tool for doing the particular industrial task.

1.2.1 MANUAL HANDLING DEVICE

A manual handling device is a manipulator under manual control or telecontrol with at least four degrees of freedom.

1.2.1.1 Manipulator under manual control

Control is exercised by direct action on the load or close to it.

Example: boom for moving load.

Exceptions: winches, hoists, rams etc.

1.2.1.2 Telemanipulator

Control is exercised at a distance using a switch-box, joysticks or a master arm.

Examples: materials handling in hostile environments: under water, in foundries, in nuclear power plants etc.

Exceptions: machines used in civil engineering and agriculture, cranes, winches, fork-lift trucks, rams etc.

1.2.2 AUTOMATIC HANDLING DEVICE

An automatic handling device is a multiaxial manipulator without servo-control but with a predetermined cycle. These machines are explained fully in Chapter 2 where *maximal effort manipulator* is the term used to describe these machines.

1.2.2.1 Automatic fixed sequence manipulator

An automatic fixed sequence manipulator is a machine that is mechanically regulated by stops and cams and is fitted with an end effector (eg gripper, spray-gun).

Examples: paint-spraying machines, press loaders and unloaders, tool loaders etc.

Exceptions: rigid conveyor or conveyor belt.

1.2.2.2 Automatic variable sequence manipulator

An automatic variable sequence manipulator is a machine which is programmable in its sequence of actions by the user, fitted with diode matrices, programmable controllers or microprocessors, and each axis is mechanically adjustable by stops and/or cams.

Examples: machine tool loading and unloading, assembly etc.

Exceptions: machine tools, transporter-stockers, motor floats etc.

1.2.3 PROGRAMMABLE SERVOED ROBOT

A programmable servoed robot is a manipulator with three or more axes under *continuous path* or *point-to-point servocontrol.* It is programmable by training (see Chapter 3), language (see Chapter 4) and computer-aided design (see Chapter 5). These machines exclude numerically controlled machine tools, machining centre etc. They are sometimes termed *playback robots.*

Subclassification:

1. Robot with three or four axes, with continuous path or point-to-point servocontrol.
2. Robot with five axes or more, with point-to-point servocontrol.
3. Robot with five axes or more, with continuous path servocontrol.

1.2.4 SECOND GENERATION ROBOT

Automatic servocontrolled manipulator able to acquire specific data concerning its work space and to react accordingly defines the second generation robot. However, reaction to binary data in the form of simple program changes is not included in this classification. Second generation robots are sometimes termed *intelligent robots*.

Examples: manipulation requiring shape recognition, assembly requiring force control, welding using seam tracking etc.

1.2.5 THIRD GENERATION ROBOT

These are characterized by their ability to sense the environment and to engage in a sequence of actions which has not been explicitly programmed. They are not seen in industry at the moment and are only experimented with in research laboratories in connection with work on artificial intelligence (eg knowledge representation, planning systems, expert systems).

Appendix I

Industrial robot: Definition for mechanics, geometry, control and programming

Translated from the Registered Standard (NF) E 61-100, August 1983

AI.1 Area of application

This standard defines the industrial robot and describes the vocabulary relating to its constituent parts and its use. As well as the robots considered in this context of this standard, the field of industrial robotics concerns other machines which are technically similar, for example:

1. programmable handling devices or manipulators;
2. telemanipulators;
3. specialized manipulators used, for example, under water, in radioactive environments or in outer space.

These mechanisms do not satisfy all the architectural or control requirements nor the characteristic or versatility demanded of industrial robots; the standard does not apply to these mechanisms, but only to those described in Section AI.2.2.

The programmable controller, defined in the Registered Standard NF C 63-850 is an electronic device for controlling machines and processes and, as such, cannot be classified as a programmable handling device in the field of industrial robotics.

AI.2 Definitions

AI.2.1 MANIPULATOR

The manipulator is made up of elements in series joined by either rotary or linear connections. Its purpose is to enable objects to be gripped and moved using a number of degrees of freedom. It is multifunctional and it can be controlled directly by the user or any logic system (eg cam system, pneumatic logic, hard-wired or programmed electric logic).

AI.2.2 INDUSTRIAL ROBOT

This is an automatic manipulator with positional servocontrol. It is reprogrammable, versatile, capable of positioning and orienting equipment, parts, tools or specialized systems in the course of programmed variable movements during the execution of varied tasks. It is often in the form of one or more arms ending in a wrist. The control unit uses a memory. In some cases it is capable of analysing the work space or environment and any changes that occur allowing the robot to react accordingly. These machines are usually required to carry out functions with a predetermined cycle but they can be adapted to other functions without permanent modification.

AI.2.3 INDUSTRIAL MOBILE ROBOT

This is an industrial robot mounted on a mobile base.

AI.3 Terminology

AI.3.1 MECHANICS

1. *Base*: the support attached to the origin of the first element in the articulated mechanical structure constituting the arm.

2. *Arm*: the chain of elements made up of linear or rotary connections which are driven in relation to each other, and carries an end effector or gripper.

3. *Wrist*: the mechanical unit at the tip of the robot arm, which is, made up of joints and allows the end effector or gripper to be oriented.

4. *Connector or coupling device for the end effector*: the device used to connect the end effector to the tip of the robot arm.

5. *End effector*: the device fixed to the wrist by the connector that performs the work for the robot (eg gripper, tool, sensor, measuring instrument).

6. *Hand*: end effector or gripper with fingers.

7. *Gripper or gripping device*: specialized end effector used for gripping and holding.

8. *Mechanical axis*: the part of the robot which can be driven with either linear or rotary movement. Mechanically, the number of axes is usually equal to the number of articulations and sliding joints driven and controlled independently.

9. *Degree of freedom* (*DOF*): one of the means, limited in number, by which a dynamic system can be modified. Each DOF is characterized by an independent variable. The number of DOF of a robot is linked to its ability to position and orient its end effector; the theoretical maximum is, therefore, six. To avoid possible confusion with the mechanical axes, the use of this expression is discouraged.

10. Slide: the mechanical unit which allows linear contact motion between two rigid segments.

11. Simple articulation: the mechanical unit which links two rigid segments and allows them relative rotary motion about a fixed axis.

12. Complex articulation: the mechanical unit which allows rotary or linear movement between two rigid parts, such that the rotation or translation is distributed or divided along the length of the unit.

13. Actuator: the power device used to drive an articulation or sliding joint.

AI.3.2 GEOMETRY

1. Pose: the position and orientation of the connector.

2. Clearance space: the volume swept by the various components of the arm during motion, which is defined relative to the base of the robot.

3. Work envelope: the volume defined by all the possible positions of the connector.

4. Effective work space: the volume defined by the positions where the performances are met for a specific application.

AI.3.3 CONTROL AND PROGRAMMING

1. Memory: the device where necessary data is stored.

2. Point-to-point control: a form of control in which the motions of the industrial robot are defined by a finite number of points through which it must pass. The position and the speed at each of these points are programmed. The different axes move either simultaneously or sequentially with respect to each other.

3. Continuous path control: a form of control in which all the positions and orientations and speed through the trajectory are defined.

4. Adaptive control: a form of control in which the parameters are determined continuously and automatically in response to variables measured as the task progresses to ensure that the task is completed efficiently.

5. Control of a mobile industrial robot: controlling the motions of a mobile industrial robot in its working environment can be carried out independent of the robot itself or by a system of input/output commands connected to the robot's own control.

6. Artificial intelligence: this is the ability to carry out functions such as reasoning, prediction, problem solving, shape recognition, perception of the working environment, recognition, comprehension and self-programming.

7. Programming: introducing into memory the necessary information for the robot to work efficiently. This can be achieved by off-line programming or programming by training.

8. Off-line programming:

Analytical programming: off-line programming by defining mathematically the trajectories and sequences.

Programming by defining the objective: off-line programming as a function of the task to be accomplished where no trajectory is imposed in advance.

9. Programming by training: execution of a model work cycle by the user. In this method, the user can introduce data into memory in different ways:

- with a manual control system characterized by the individual control of each actuator;
- with a device allowing intuitive and overall manual control of the position and orientation of the end effector of the robot;
- by direct action on the tip of the robot;
- with a device similar in shape to the robot, with which it is exchanged. Despite the fact that the device is lighter than the robot the user's movements are not affected or modified.

Appendix II

Industrial robot: Designation of geometrical axes and movements

Translated from the Registered Standard (NF) E 61-101, August, 1983

AII.1 Area of application

This standard fixes the geometrical definitions of the mobile parts of an industrial robot as well as the terminology of the movements at the tip.

AII.2 Reference

See NF E 61-100.

AII.3 Geometrical reference position

The axes of motion of an industrial robot are defined by an orthogonal trihedron with Cartesian coordinates T_0, horizontal axes X_0 and Y_0 and a vertical axis Z_0. These axes which characterize the mechanical structure are present in a position such that:

- the base is in the plane X_0Y_0 of the *reference trihedron* with origin O_0 where the first element of the arm is attached to the base;
- all the axes of motion are parallel to the directions of the axes of the reference trihedron or *reference set of axes.*

This position is called the *geometrical reference position.*

AII.4 Geometrical axes associated with the mobile parts of the industrial robot

In order to characterize an industrial robot the geometry and the terms specific to the movements of the end effector should be explained.

AII.4.1 GEOMETRY EXPLAINED

The functional geometrical definitions are intended to describe the spatial position and orientation of the tip of the arm of the industrial robot, and also intermediate moving parts or segments. The positions of the mobile parts of the industrial robot are defined in the reference trihedron $T_0(X_0, Y_0, Z_0)$ with origin O_0 on the base of the robot. The moving parts of the industrial robot are numbered from 1 to n, from the base to the tip, each associated with trihedron $T_i(X_i, Y_i, Z_i)$ with origin O_i on the rigid part of the mobile element i and on the common part (eg axis, slide) with elements i and i-1. The axes of the trihedron T_i are aligned with the mechanical parts (eg axes, joints, slides) linked to the moving part i. The robot is said to be in reference position when the trihedrons T_i have a fixed identified orientation, known as the *reference orientation* relative to trihedron T_0. A trihedron $T_e(X_e, Y_e, Z_e)$ parallel to trihedron T_n is associated with the tip of the industrial robot arm. Its origin O_e is situated at the tip of the last moving part of the arm, that is the connector or coupling device. The position of the tip of the arm is given by the coordinates of O_e in the trihedron T_0. The orientation is defined by the three angles:

- λ, yaw;
- θ, pitch;
- ρ, roll.

Figure AII.1 shows orientation of the tip trihedron $T_e(X_e, Y_e, Z_e)$.

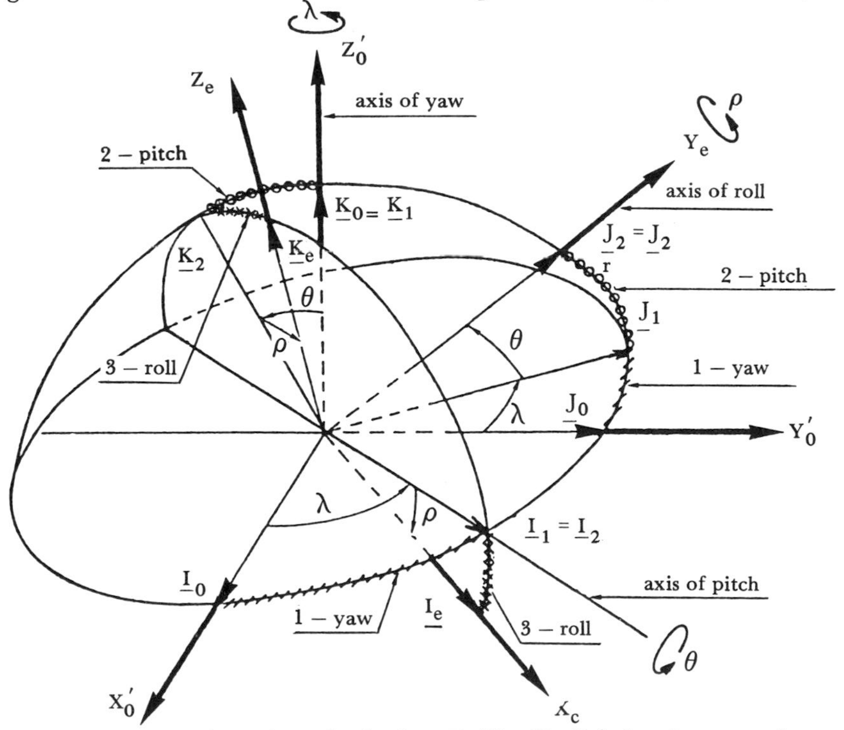

Figure AII.1. *Orientation of trihedron $T_e(X_e, Y_e, Z_e)$ showing axes of yaw, pitch and roll*

The different rotations which allow the tip trihedron to be oriented relative to the base trihedron are:

1. Yaw Rotation of axis $\underline{K_n}$: λ
   $\underline{I_0}\underline{J_0}\underline{K_0}$ $\underline{I_1}\underline{J_1}\underline{K_1}$
2. Pitch Rotation of axis $\underline{I_1}$: θ
   $\underline{I_1}\underline{J_1}\underline{K_1}$ $\underline{I_2}\underline{J_2}\underline{K_2}$
3. Roll Rotation of axis $\underline{J_2}$: ρ
   $\underline{I_2}\underline{J_2}\underline{K_2}$ $\underline{I_e}\underline{J_e}\underline{K_e}$

The matrix for change of coordinates (P_a) is:

$$\begin{bmatrix} X'_0 \\ Y'_0 \\ Z'_0 \end{bmatrix} = \{P_a\} \begin{bmatrix} X_e \\ Y_e \\ Z_e \end{bmatrix} \quad \textbf{(AII-1)}$$

$$\{P_a\} = \begin{bmatrix} \cos\lambda\cos\rho - \sin\lambda\sin\theta\sin\rho & -\sin\lambda\cos\theta & \cos\lambda\sin\rho + \sin\lambda\sin\theta\cos\rho \\ \sin\lambda\cos\rho + \cos\lambda\sin\theta\sin\rho & \cos\lambda\cos\theta & \sin\lambda\sin\rho - \cos\lambda\sin\theta\cos\rho \\ -\sin\rho\cos\theta & \sin\theta & \cos\theta\cos\rho \end{bmatrix} \quad \textbf{(AII-2)}$$

AII.4.2 DEFINITIONS SPECIFIC TO MOVEMENTS OF THE END EFFECTOR

The movements at the tip of the arm are defined by translations and rotations; they can be expressed as:
In trihedron $T_0(X_0, Y_0, Z_0)$:

- a translation along axis X_0 by left-right;
- a translation along axis Y_0 by front-back;
- a translation along axis Z_0 by up-down.

and in trihedron $T'_0(X'_0, Y'_0, Z'_0)$, which is a trihedron parallel to T_0, with origin O'_0 is confused with O_e:

- a rotation about axis X'_0 by site;
- a rotation about axis Y'_0 by warp;
- a rotation about axis Z'_0 by azimuth.

In trihedron $T_e(X_e, Y_e, Z_e)$:

- a translation along axis X_e by sweeping;
- a translation along axis Y_e by extending and retracting;
- a translation along axis Z_e by lifting and lowering;
- a rotation about axis X_e by inclination;
- a rotation about axis Y_e by torsion;
- a rotation about axis Z_e by rotation.

Notes

In trihedron T'_0:

- when λ and ρ are simultaneously zero, a site rotation is a pitching rotation;
- when θ and λ are simultaneously zero, a warp rotation is a rolling rotation;
- an azimuth rotation is always a yawing rotation.

In trihedron T_e:

- when λ and ρ are simultaneously zero, an inclination rotation is always a pitching rotation;
- when θ and ρ are simultaneously zero, a rotation is always a yawing rotation;
- a torsion rotation is always a rolling rotation.

AII.5 Orientation of the wrist

The orientation of the end effector in relation to the wrist is defined by the orientation of the trihedron T_e in relation to the intermediate trihedron T_k anterior to the wrist. This capacity for orientation is generally independent of the position of the tip of the industrial robot arm.

AII.6 Orientation of the end effector or gripper

The orientation of the terminal part is characterized by the angles γ, θ, and ρ. These angles are dependent on the position of the tip of the industrial robot arm.

Chapter 2

Maximal effort manipulators

2.1 Maximal effort manipulators or pick-and-place robots?

A *maximal effort manipulator* is a multiaxial mechanical unit in which the end effector is capable of occupying only a limited number of positions whilst maintaining steady operation. This form of control is termed *bang-bang control.* These robots are sometimes called *pick-and-place robots* but pick-and-place robots can sometimes include servo-driven robots (see Chapter 3). If all axes are under bang-bang control the space to be reached by the end effector (known as the work space or envelope) is limited to a finite set of points. The number of stable points per axis is for technical reasons generally two. Thus, the total number of attainable points is 2^n if n is the number of axes.

Example: a robot with three axes with cylindrical coordinates has $2^3 = 8$ stable points (see Figure 2.1.a).

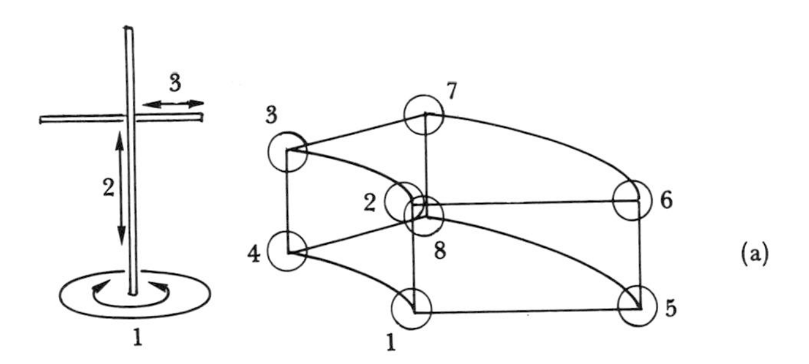

Figure 2.1. *(a) Work envelope defined by eight points*

If one of the axes is subjected to positional servocontrol, the work space is made up of a finite set of curves in the space (2^{n-1}). Therefore, since the rotational axis 1 is servocontrolled, the work space is made up of four curves a, b, c and d (see Figure 2.1.b):

$$2^2 = 4 \text{ curves}$$

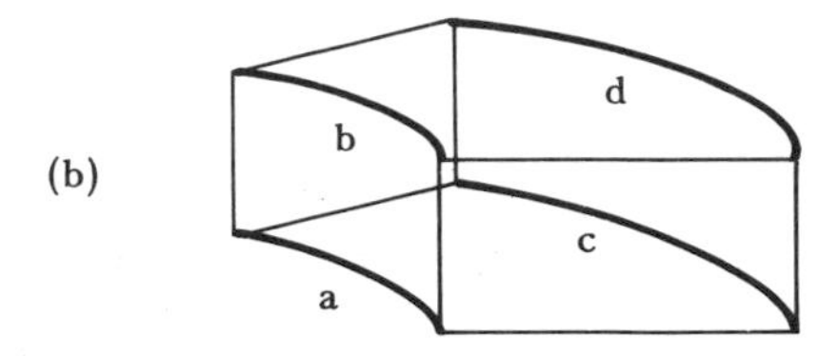

Figure 2.1. *(b) Work envelope defined by four curves a, b, c and d*

If two axes are subjected to positional servocontrol, the work space is made up of a finite set of surfaces (2^{n-2}).

In Figure 2.1(c) the work space is made up of two surfaces S_1 and S_2, when axes 1 and 2 are servocontrolled.

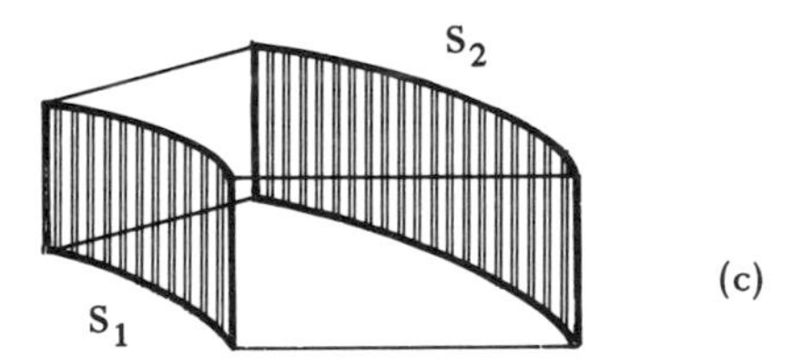

Figure 2.1. *(c) Work envelope defined by two surfaces S_1 and S_2*

If a third axis is subjected to servocontrol, the work space is made up of a finite set of volumes.

Thus a situation may arise in which a given manipulator is neither entirely bang-bang controlled nor entirely servocontrolled. This results in a hybrid mechanism with both continuous path and bang-bang control. All the axes of a robot may be servocontrolled, but if the command points of the servo-system are managed using sequential logic, the robot is described as a *pick-and-place manipulator.* The distinction is not always clear, and arises from the problem of discrete versus continuous motion.

Consider a robot with n axes, each one with m stable attainable positions, and p is the number of steps or number of configurations that can be memorized for each axis. The number of points that the end effector can reach is m^n. The number of configurations of the end effector that can be memorized during the cycle is p^n. Consider the example of an arm with three axes, programmed using a diode matrix for 20 steps, with 10 stable positions per axis.

$$n = 3 \qquad m = 10 \qquad p = 20$$

Thus $m^n = 10^3 = 1{,}000$ attainable points and $p^n = 8{,}000$ memorized configurations. A robot of this type would, nonetheless, be categorized as a pick-and-place manipulator.

Analog recording of command trajectories theoretically allows an infinite number of possible configurations. This number is, in practice, limited by the precision of the servo-system. This technology which is adopted by *playback robots* is becoming less and less widely used. The widespread availability of inexpensive memory systems has allowed digitally controlled robots to be profitably used in industry.

2.2. Importance of maximal effort robots

Although robots from the upper end of the price range are still generally out of reach of most small- and medium-sized industries, *fixed or variable sequence handling devices* can generally pay for themselves within eight months of purchase. The number of potential applications is enormous. The importance of maximal effort robots is the result of a combination of the low cost of the parts and the many applications to which these robots can be put.

These applications concern any handling operations in which only the starting and finishing point needs to be located precisely:

1. arms for loading or unloading transfer machines or machine tools;
2. transfer between work points or workstations;
3. loading and unloading machines (eg lathes, pressure moulds, injection moulds);
4. storing and palletization;
5. automation of assembly, welding, cutting devices etc.

The reasonable cost of the equipment results from the fact that a large majority of maximal effort robots use pneumatic actuators. Pick-and-place robots which use hydraulically driven systems are a great deal more expensive.

The number of axes or degrees of freedom (DOF) can vary between two and five, and can be either rotational or translational. These robots are used for repetitive work. The cycle is the set of elementary actions defining one of the single operations which is repeated indefinitely. During a single cycle, a given axis may be driven once or several times. Let a^+ and a^- be the outward and return strokes of actuator A. With a single axis, the cycle can only be as follows:

$$(a^+a^-) \text{ or } (a^+a^-a^+a^-) \ldots (a^+a^- \text{ repeated n times } a^+a^-)$$

The simple cycle (a^+a^-) is called a *pendular cycle*, and as all other possible cycles are repetitions of this cycle, a counting mechanism is required.

If two actuators A and B are used, the following cycles can be produced:

$(a^+b^+a^-b^-)$ which is called a *square cycle*, and is purely combinational from a logic point of view and
$(a^+b^+b^-a^-)$ which is called an *L cycle* and involves sequential logic.

For example, if a drill is to be moved so as to pierce an object at four points forming a rectangle, a square cycle will be used (see Figure 2.2.a), thus:

$$(a^+b^+a^-b^-)$$

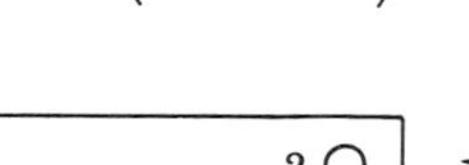

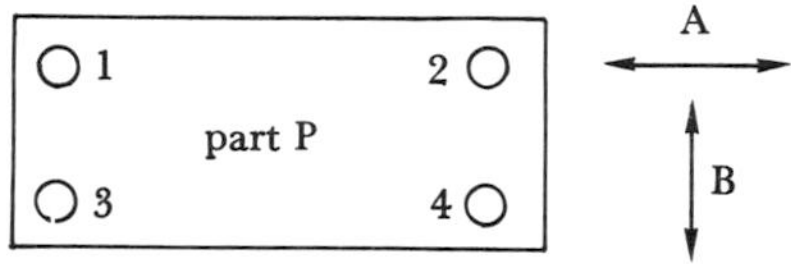

Figure 2.2. *(a) When a drill is moved to punch four holes forming a rectangle in a plate a square cycle is used*

If, however, a drill F is to be used to pierce an object wedged in place by a gripping device A, an L cycle will be used (see Figure 2.2.b), thus:

$$(a^+f^+f^-a^-)$$

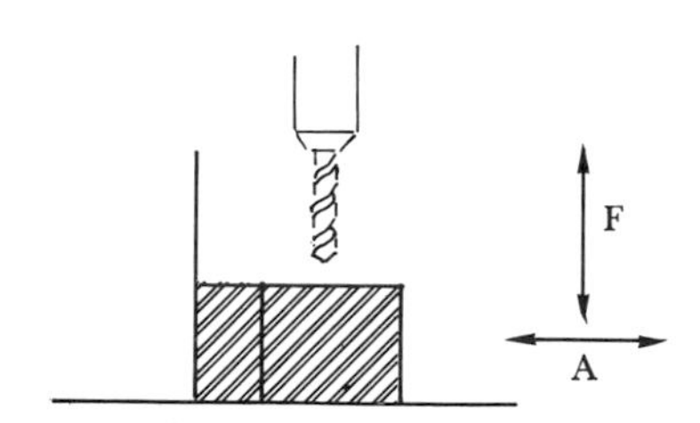

Figure 2.2. *(b) When a drill (F) is used to pierce a plate wedged in place by the gripper (A) an L cycle is used*

All two-axis cycles arise from different arrangements of these two basic cycles. For example, if the drilling task involves chip removal, actuator F will be activated a second time and the cycle will become:

$$(a^+f^+f^-f^+f^-a^-)$$

If a robot with three axes is to be used, the description of the cycle can also be either completely *combinational*:

$$(a^+b^+c^+a^-b^-c^-)$$

or *sequential*:

$$(a^+b^+c^+c^-b^-a^-)$$

The various cycles can be visualized as movements along the edges of a cube. In all cases, the cycle must end at the point where it started (vertex A of the cube) (see Figure 2.3).

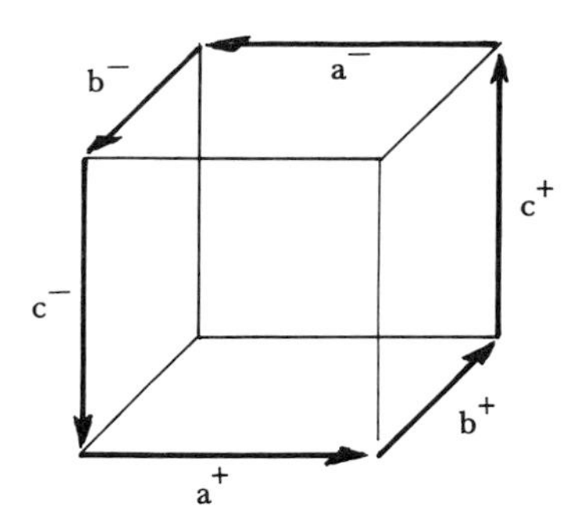

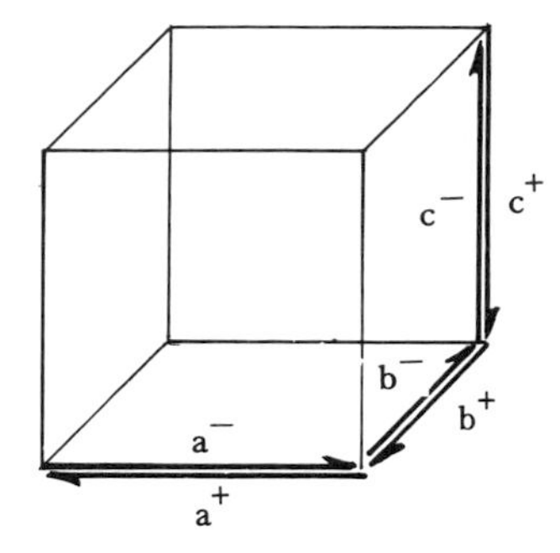

Figure 2.3. *Diagrams showing combinational and sequential cycles imposed on a manipulator with three axes (three DOF)*

With four or five axes or more the cycle can be described in the same way. The chain can, however, contain subsets made up of subcycles repeated a given number of times. Not all the actions are necessarily sequential over a given period, and some may be carried out in parallel, thus involving synchronization mechanisms.

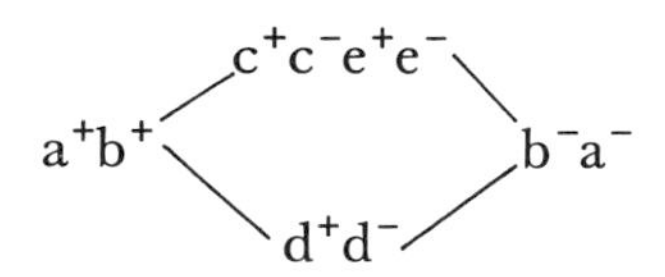

Thus it is necessary to use a method to describe these functioning cycles that transfers easily to the programming of pick-and-place robots. Although there are many methods available, the only one considered here is *Grafcet.*

2.3 Grafcet and binary memories

2.3.1 GRAFCET

Grafcet is a way of describing the specification of a logic automaton, and as such is used to describe the operation of pick-and-place robots. This method, devised in France, has achieved considerable popularity in a relatively short time. The main concepts and rules for its development in the control of robotic systems will be discussed in the appendix to this chapter.

If a layman attempts to analyse the operation of an automated machine, he perceives the sequential mechanism in terms of reasoning of the kind: 'I press the start button, the motor starts up, it reaches a stop which activates a pump etc'. If the description consists simply of n elements of binary information on the automated system, this leads to 2^n different configurations. The human brain rapidly loses track of all the possibilities, which are many. The vast majority of these configurations are of no use with respect to the operation of the automated system. The overall idea of Grafcet is to isolate the viable combinations for developing the system from the vast number of subsets presented.

The success of Grafcet because of its simplicity and its educational value allows automatons to be fully understood after a short instruction period. Grafcet is not, however, the only method with varying degrees of specificity and varying types of application. There are at least 20 other methods available, including *ladder diagrams*, *relay networks, Huffman method, Climax method, Petri nets* etc. Some of these methods are discussed in relation to specific examples, showing their importance and suitability to certain types of application.

2.3.2 BINARY MEMORIES

In speaking of a robot controlled by a processing device such as a mini- or microcomputer, it is normal to refer to the robot *programming language.* Yet it is clear that in dealing with such a programmable controller, the language is specific to the controller rather than to the robot. Specifically robotic languages are related to software on general purpose computers, and can be described as *problem-oriented languages.* In other words, the control unit of the robot can be described as a *language machine.* On the other hand, for pick-and-place robots controlled by electric or pneumatic hard-wired logic, the term *programming language* is inaccurate. The language must be seen as the means of expressing to a *control unit* the operations the automaton must carry out. For processing devices, this takes the form of machine code; for hard-wired logic, the language is the method of associating the technological elements. In hard-wired logic, however, the controller is not an arithmetic and logic unit, but it is based on binary memory units.

A *binary memory* is a technological element equipped with an output (and often a second complementary output) and two binary inputs, and in which it is possible to read, write and clear the binary information of the output variable. One of the inputs is used to write to the memory ($S = Set$). The second input is used to clear the memory ($R = Reset$). Out of the four possible inputs (00, 10, 01 and 11), two are used to write to and clear the memory (10, 01), one is used for memorization (00) and the fourth combination (11) allows five types of memory to be discerned: *priority set, priority reset, RS bi-stable, JK bi-stable* and *indifferent memory* (see Table 2.1).

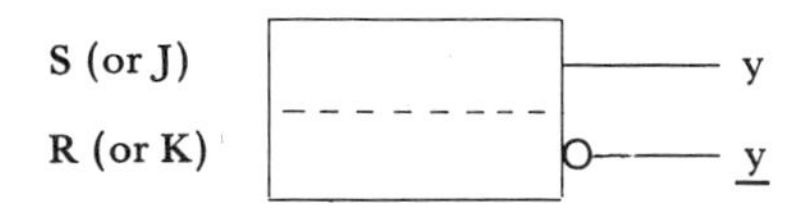

SR \ Y_{n+1}	priority set memory	priority reset memory	RS bi-stable	JK bi-stable	indifferent memory
00	Y_n	Y_n	Y_n	Y_n	Y_n
10	1	1	1	1	1
01	0	0	0	0	0
11	0	1	?	$\underline{Y}_n$	Y_n

Table 2.1. *Y_{n+1} represents the output at time T_{n+1} after a combinational SR has been applied to the memory at time T_n when the output was Y_n*

2.4 Electromechanical relay logic

Many of the transfer robots with two or three axes used in industry are still controlled by relays. Two examples are:

1. Robonorm by Automatisme Européen;
2. Manumax by Climax.

Programming is based on the method of wiring the relays. Consider a two-axis transfer arm which must grip objects at P_1 and move them to P_2. The cycle chosen is:

$$(a^+b^+a^-b^-)$$

The following Grafcet can be written (see Figure 2.4).

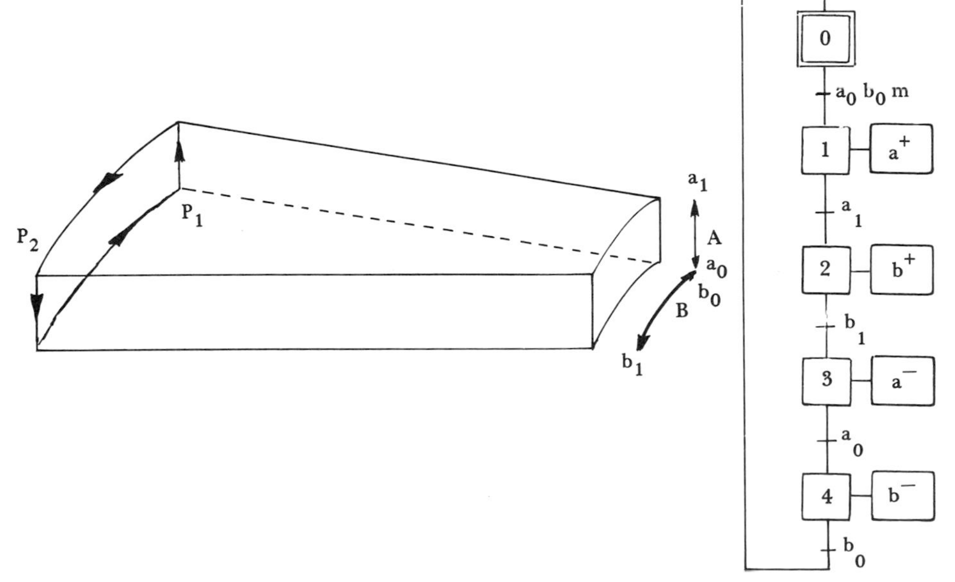

Figure 2.4. *Cycle of a transfer arm and associated Grafcet*

The application of Grafcet leads to a sequential process which involves five binary memories associated with five levels, whereas the problem is in fact far simpler because it is combinational. A place diagram or a Karnaugh table provides the simple solution shown in Figure 2.5.

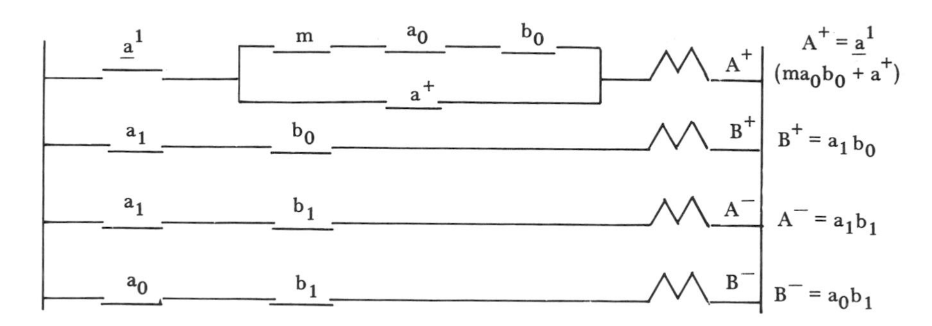

Figure 2.5. *Electrical schema and Boolean equations*

This elementary example shows that simple relay technology as well as older methods (eg Karnaugh table, Huffman method, Place diagram) can be justified for use with small systems. The relay industry is generally conservative and thus retains a significant number of older approaches, although development is proceeding:

1. modules of the same size as relays are being developed with more complex functions (eg RH sequencer modules);
2. relays which can be directly interfaced with the control electronics are being designed;
3. low-power relays are being miniaturized to ensure mechanical compatibility with printed circuits;
4. electronic functions such as timing, counting and buffering are being integrated with regard to relays.

At this point, static electronics should be mentioned. In relays the moving parts have a response time in the order of a millisecond. Static relays, whose name was derived from the fact that there are no moving mechanical parts, are the same size as traditional relays but electronic commutation allows the command signal to be smaller and the response time to be far shorter. Despite a successful launch, however, this technology has not presented any real competition to electromagnetic relays.

2.5 Pneumatic logic

Since 1976, the year the first sequencers were produced, pneumatic logic has enjoyed considerable success, to such an extent that although in 1972 only 5 per cent of automated systems were pneumatically driven, by 1977 20 per cent used the technology, and the figure is expected to stabilize at the 50 per cent mark. Its success is the result of a number of technical factors:

1. actuators (rotary or linear pistons), sensors and control now rely on the same technology;
2. popularity of pneumatic modular and juxtaposable sequencers;
3. high level of similarity to electronic components (eg profiles, plates, frames);
4. easy repair by visualization of cycles;
5. safe use in explosive or dangerous areas;
6. complete insensitivity to radio-electric fields;
7. very low level of heat dissipation relative to power levels used;
8. standardized connections.

There are many methods by which pneumatic logic can be programmed, including the Climax method, extended Karnaugh tables, the geometrical method, the cascade method etc. Often the method chosen is determined by the technical elements. The piston may have either single or double action, the distributors may be either pneumatically or electrically controlled (electrodistributors) or spring-loaded.

Many methods implicitly take into account the fact that the 'two track, five aperture distributor', either on its own or associated with its actuator, constitutes a *pneumatic memory*, and behaves as an RS bi-stable (see Figure 2.6).

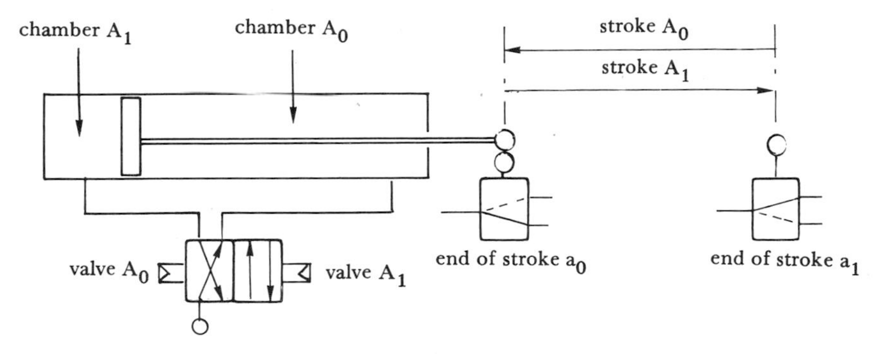

Figure 2.6. *Actuator-sensor-distribution assembly*

The activation of A_1 brings about an outward movement by the actuator, thus placing the sensor a_1 at 1 (set in memory). In the same way, the activation of A_0 causes the piston to return (clearing the memory). Because of the spool distributor, the commands at A_0 and A_1 are pulsed and do not need to be maintained (memory function). On the other hand, A_1 and A_0 should not be activated simultaneously, as for the RS bi-stable.

2.5.1 THE CASCADE METHOD OR LADDER DIAGRAMS

Mention has been made of the square cycle and the L cycle. If these cycles are represented geometrically, it can be seen that for a single distribution of the sensor inputs in a_1b_0, the L cycle leads to two types of decision B_1 or A_0. In other words, in the square (combinational) cycle knowledge of the input configuration allows the decision to be taken regarding the output to be activated. On the other hand, in the L cycle (sequential automaton) each input configuration does not always correspond to a unique output configuration (see Figure 2.7).

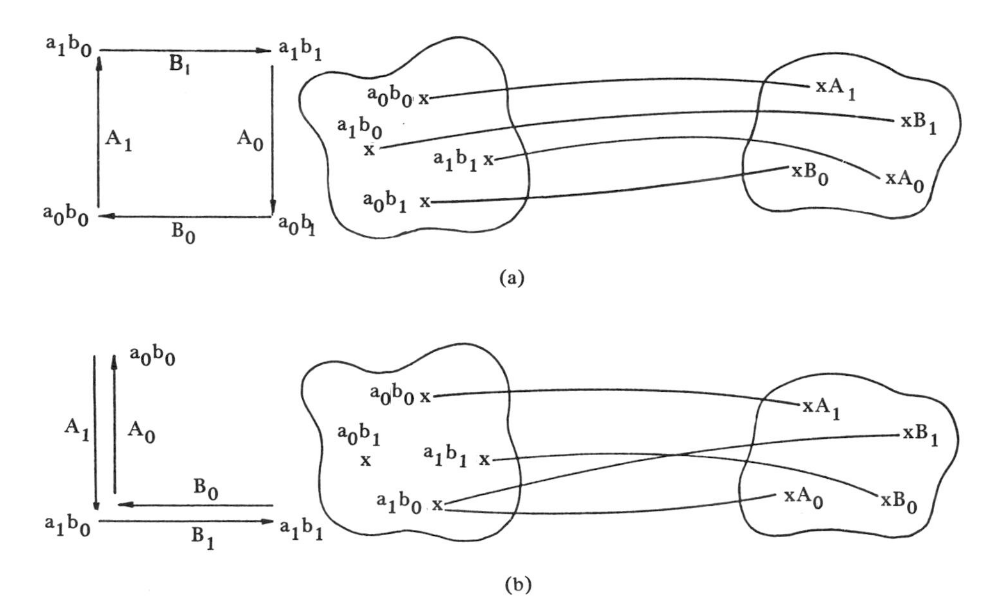

Figure 2.7. *Scheme showing combinational and sequential automatons: (a) combinational square cycle; (b) sequential L cycle*

Programming combinational automatons is particularly simple because any Boolean equation associated with an input configuration brings about the activation of a single output. The idea behind the cascade method is to break down, if necessary, any cycle into several combinational subsets.

In the description of $a^+b^+c^+a^-b^-c^-$ type cycles, the representation is devised in sections so that each actuator appears only once in each section.

Example: $a_1b_1c_1b_0c_0a_0$ becomes $a_1b_1c_1/b_0c_0a_0$. This ensures that each section is combinational, and thus a binary memory is associated with each section. The basis of the cascade method will be explained in connection with this cycle.

1. The cycle is divided into two sections, each containing an actuator once. The cycle is thus made up of two sections that can be distinguished by a memory x and a memory $\underline{x}$:

$$\frac{A_1B_1C_1}{\overline{x}} \Big/ \frac{B_0C_0A_0}{x}$$

2. A program is established:

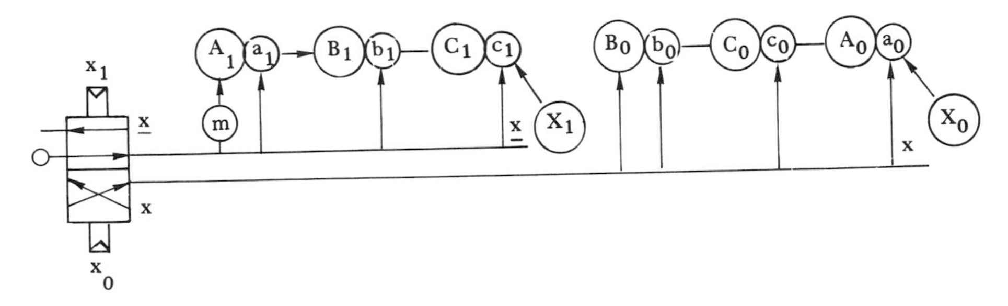

This program reads: 'The activation of A_1 by the supply line $\underline{x}$ via switch m mechanically activates sensor a_1, which supplies B_1 which drives b_1 . . . '. At each end of each section, the memory is activated or deactivated (in this case the memory is x). The limits of the section strokes and the first value of the section are supplied by the line associated with the section.

3. Boolean equations are obtained directly from the program:

$$\begin{aligned} A_1 &= m\overline{X} & \qquad A_0 &= C_0X \\ B_1 &= a_1\overline{x} & B_0 &= x \\ C_1 &= b_1\overline{x} & C_0 &= b_0X \\ X_1 &= c_1\overline{x} & X_0 &= a_0X \end{aligned} \qquad (2\text{-}1)$$

4. The developed schema of the installation is thus immediate and the reader can follow the development of the cycle (see Figure 2.8). It is clear that such an effective method relies on the technology of the pneumatic components and cannot be applied to an electrical system. It should be noted that the command logic is carried by the distributors and sensors at the end of the strokes, which makes the set of pneumatic components homogeneous.

2.5.2 SEQUENCERS

Sequencers are modular logic control units, and can be electromechanical, static electronic and, above all, pneumatic. The basic module is formed from a binary memory, an AND logic cell and an OR logic cell.

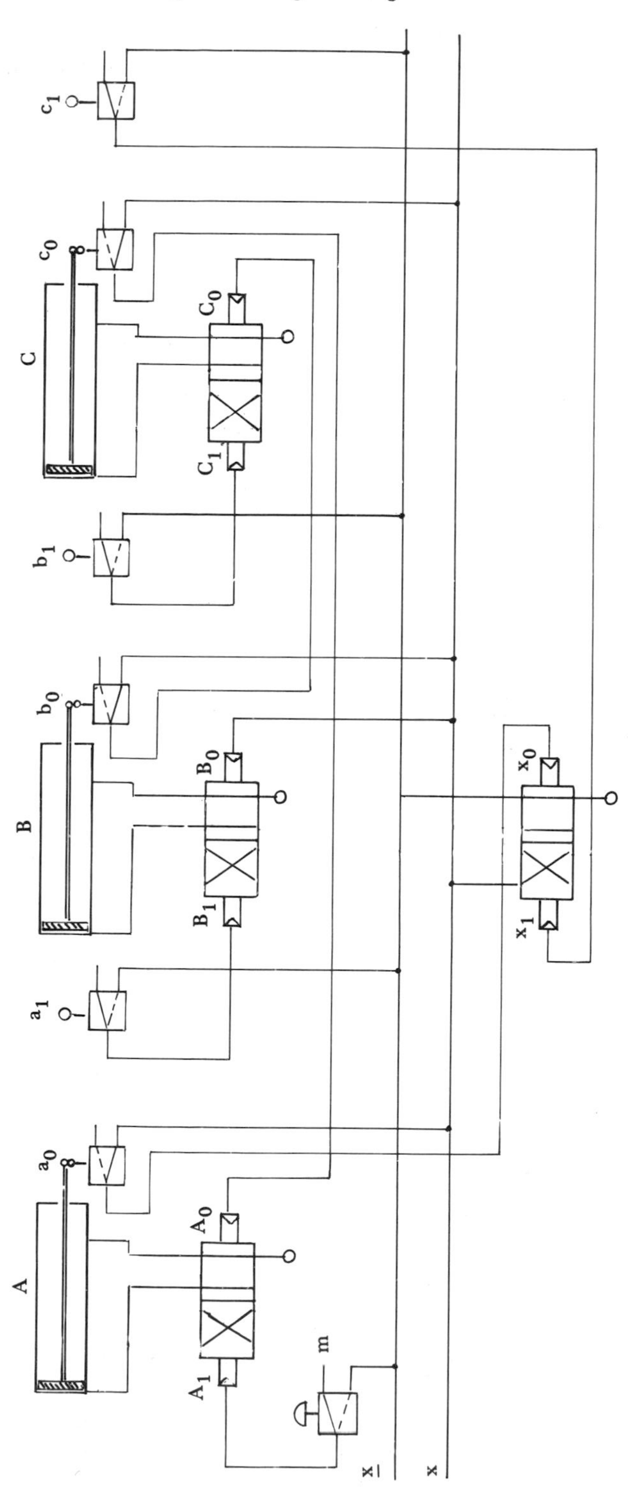

Figure 2.8. *Pneumatic cabling of the A_1 B_1 C_1 B_0 C_0 A_0 cycle*

The sequencer is formed by the association, linear or not, of modules. The number of modules is equal to the number of steps in the cycle to be carried out.

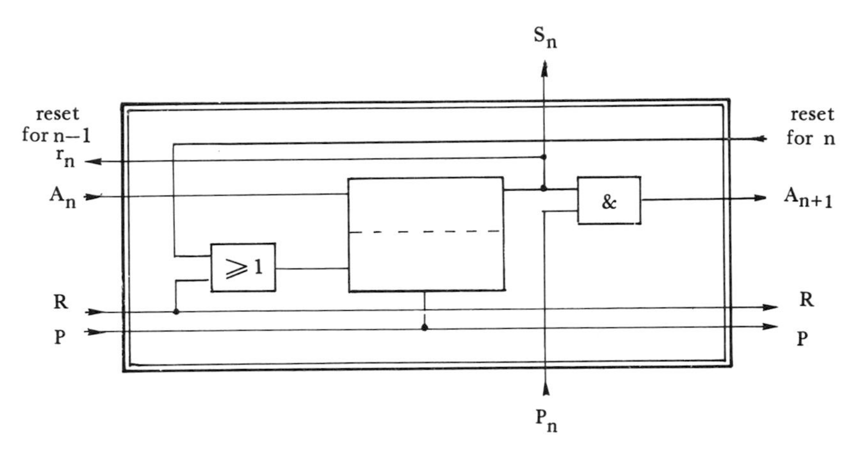

Figure 2.9. *Scheme showing the principle of the sequencer*

In Figure 2.9, P represents the pressurized supply to the module, R is a signal for return to zero, acting by the intermediate action of an OR function, A_n is the activation of module n coming from module $(n-1)$, r_n is the feedback of information consecutive to the action brought about at S_n. Connecting the modules together on a profile ensures connection continuity, and thus minimizes the hard-wiring of the logic functions. An analogy can be made by considering the sequencer module as an integrated pneumatic module and the sequencer itself as a printed circuit. This technique of forming a stepping device is a relatively recent development since the first sequencers appeared in 1976. This is the result of production problems.

2.5.3 EXAMPLE: TRANSFER OF PARTS BETWEEN TWO CONVEYOR BELTS

A pick-and-place robot is required to transfer plates from conveyor belt A to conveyor belt C (see Figure 2.10). The working part of the system is made up of four actuators (therefore eight possible outputs to be activated) and seven binary sensors. The representation of the cycle, counting of the inputs and outputs, as well as Grafcet allow the user to understand thc operation of the automaton without additional information (see Figure 2.11).

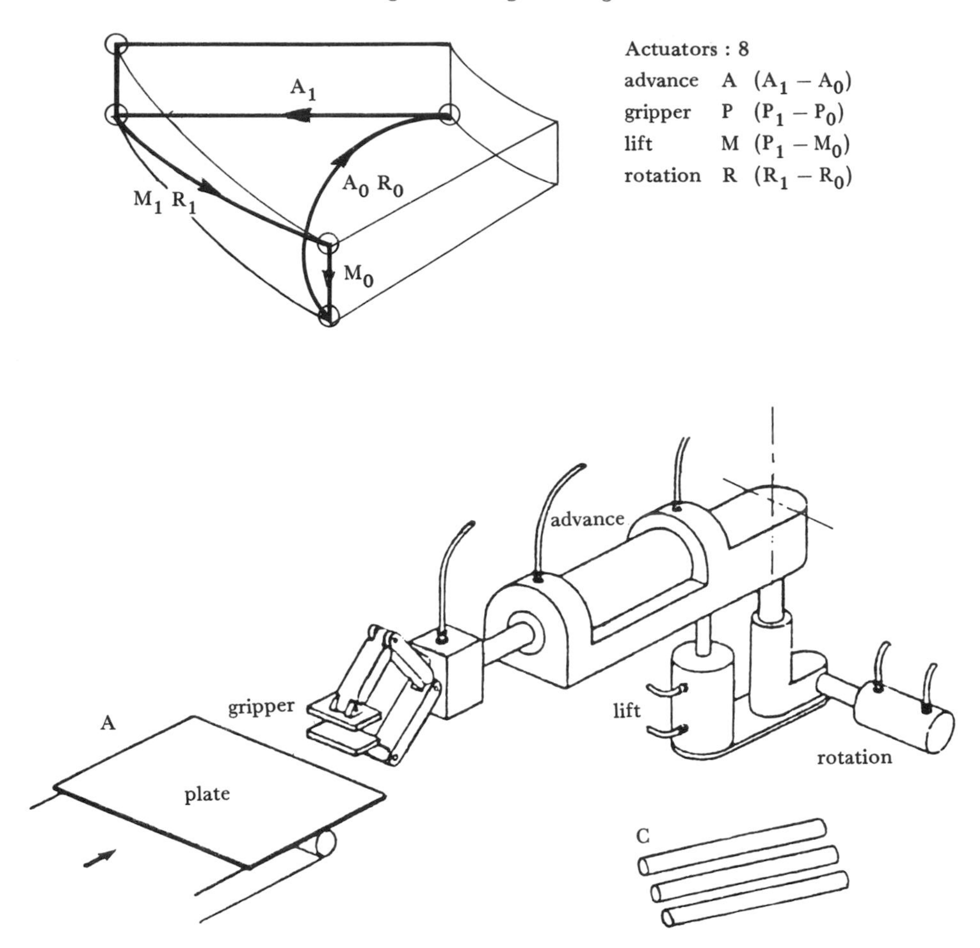

Figure 2.10. *Scheme showing transfer point*

Only six modules are required to hard-wire this cycle on a Crouzet-type sequencer (see Figure 2.12).

Switch m is used to activate module 1, and so to start the cycle. Some actions are carried out in parallel to save time (and so to increase productivity), for example M and R. Parallelism can be achieved using a number of solutions; the one shown below interlocks the two actions, and the slower of the two marks the end of the place. If the action is independent, for example a piston shaft making an outward movement to its limit, this solution can be used. If, on the other hand, the action cannot tolerate the receptivity (see Appendix) resulting from the parallelism, for example a rise in temperature or an increase in pressure, the solution would be impractical and possibly dangerous. A better solution in such a situation is to make the two actions completely independent of each other, with synchronization at the beginning and

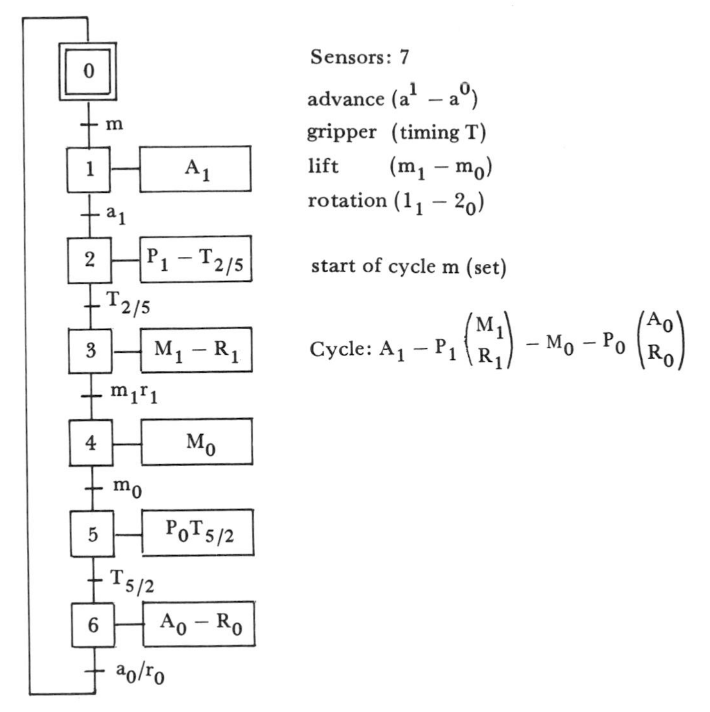

Figure 2.11. *Analysis of a geometric cycle and its corresponding Grafcet*

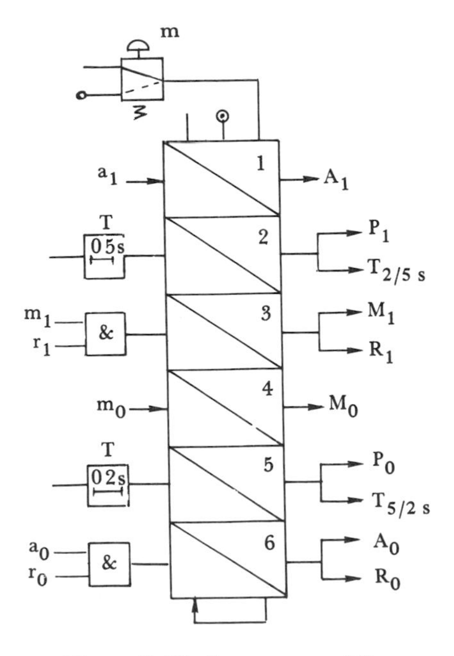

Figure 2.12. *Sequencer cabling*

at the end. This solution is possible using Grafcet, with translation onto a sequencer, as shown below (see Figure 2.13). In the same way, sequencers can be used to carry out direct shunting, repetition and jumps forward or backward.

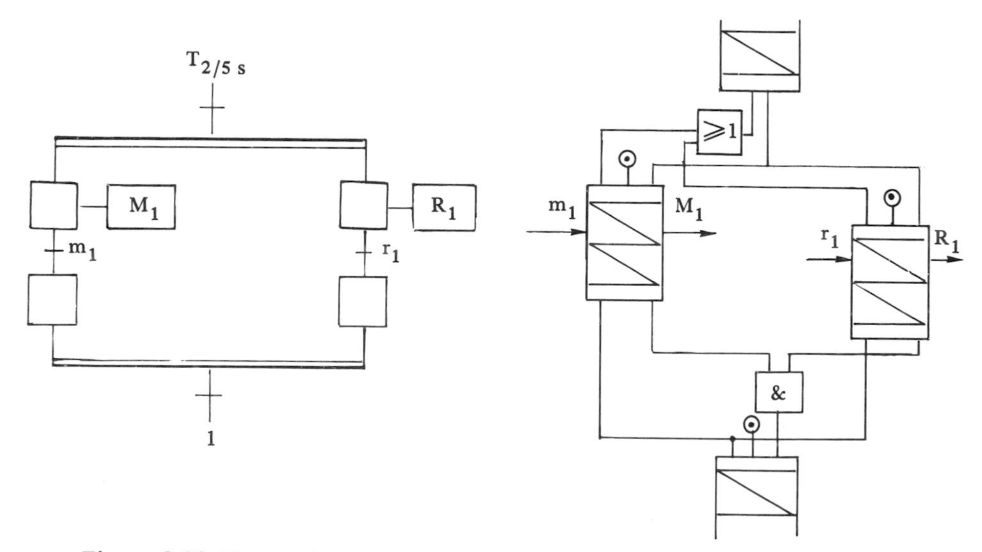

Figure 2.13. *Two actions completely independent of each other with initial and final synchronization*

Each module in the sequencer has a control that can be used to verify the active or inactive state of the module. This facility is extremely useful when looking for breakdowns and establishing installations.

A module costs about $15. Therefore, it is not surprising that this technology, which is suitable for small systems, inexpensive, reliable, easy to use and repair, and completely insensitive to noise, has achieved considerable success.

2.6 Programmable controllers

2.6.1 MAIN CONCEPTS

Programmable controllers appeared in the USA in about 1970. *Programmable controllers* can be defined as 'programmable machines that can be used by the non-specialist user, and are intended to establish simple, logic or binary automatons in real time in an industrial context'. The significant success of these devices is derived from a number of factors:

1. They are robust mechanisms which are technically designed for industry.
2. They are language machines. Computers, microprocessors and programmable controllers respectively have basic instruction sets of the following orders of magnitude:

– computers	250 instructions
– microprocessors	60 instructions
– programmable controllers	20 instructions

   It may seem strange that machines with such limited potential as programmable controllers can compete with general purpose computers. This is a result of the careful choice of very powerful instructions, oriented towards automata, which leads to machine code or problem-oriented language. The timing instructions, for example, are extremely powerful and can be used to cut-off and control the time axis. The memory instructions, counting, sequencers, master relays etc are also examples of *primitives* in a language oriented towards controllers and are absent in general purpose computers.
3. They are machines which separate the production and execution functions in the program, using the console (see Figure 2.14).

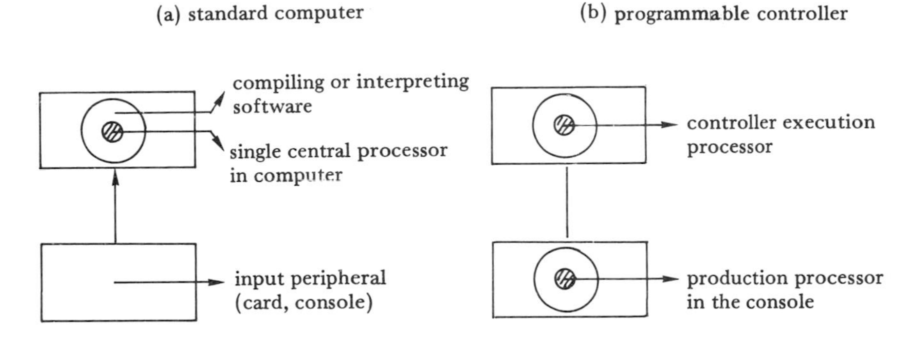

Figure 2.14. *Principle of the monoprocessor standard computer (a) and the biprocessor programmable controller (b)*

In a standard computer the translation (from high-level language to binary code) which constitutes the production phase of the program is carried out by the processor, which then uses the binary code. However, in the programmable controller, on the other hand, the two phases are separated. Production of the program is carried out by the console processor and execution by the controller itself. It is in this sense that the programmable controller can be described as a dual-processor machine. Moreover, the translation of high-level language into binary code is bi-directional, which transforms the console into a powerful tool for adjustment and repair. This physical separation of the production and execution functions provides a high degree of flexibility by allowing off-site programming and debugging.

4. They are machines which process real-time problems using a synchronous approach. In general, these machines cannot be interrupted, but are equipped with an implicit cycle of a few milliseconds. Any action is therefore processed with a delay by the processor involved. The maximum length of the delay will be equal to the length of the cycle. The real-time approach of industrial computers, on the other hand, is generally asynchronous in that the interruptible processor gives precedence to the most urgent tasks at hand. This basic difference is not determinant in itself, but it is this implicit cycle that allows simple programming of timing problems by a user not trained in computing.

2.6.2 SOFTWARE FOR PROGRAMMABLE CONTROLLERS

The various types of software can be considered in terms of an analysis of their functions:

1. logic function;
2. numerical calculation function;
3. regulation function.

The regulation function can be omitted, since the calculations involved take too long, and so rules out the use of programmable controllers with mechanical servo-systems for robots. The regulation functions are more useful in the control of slow industrial processes.

The logic and numerical calculation functions, on the other hand, can make the programmable controller a tool particularly well suited to controlling maximal effort handling devices. The software available for Boolean calculation falls into two categories:

1. The American school, which makes use of relay language and graphic symbols, in the form of ladder diagrams. Figure 2.15 shows the keyboard of a console indicating the primitives of this language and the meaning of the main function keys.
2. The European school, which makes use of a Boolean or numerical-type representation.

In the Boolean system described as *horizontal*, the Boolean equation is introduced directly in its written form at the programming console. The connection with the relay circuit and the logic equation which constitutes a macro-instruction is shown below (see Figure 2.16).

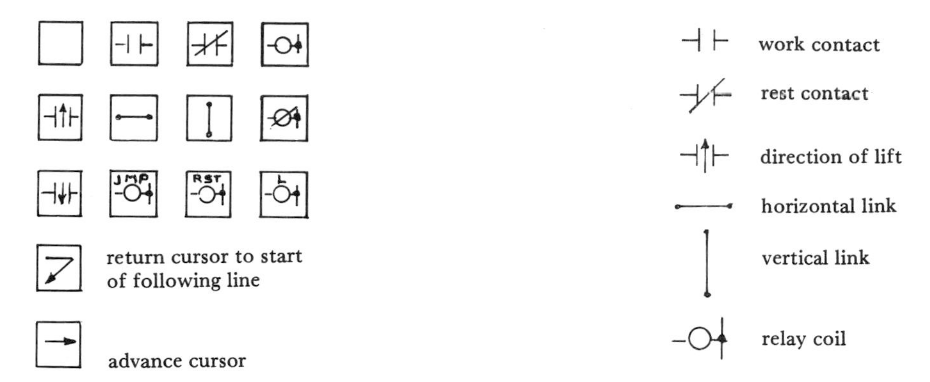

Figure 2.15. *Main function keys represented by graphic symbols*

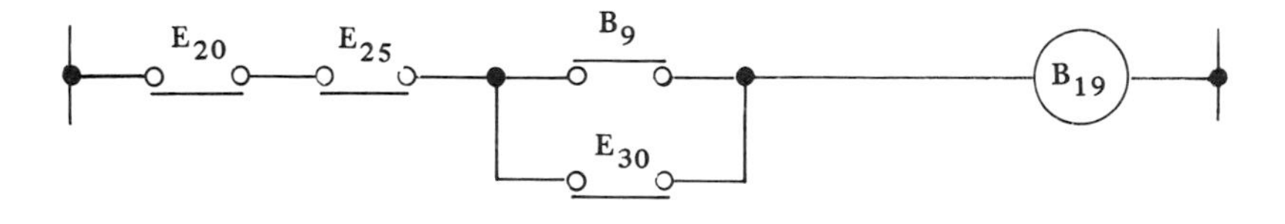

S10.E20.E25 [/B9 + E30] = B19. → validation

equation

→ line validation

→ sequence identification

→ result of equation

Figure 2.16. *Relay circuit (a) and the corresponding Boolean macro-instruction (b)*

In the Boolean system described as *vertical*, the equations are translated using assembler-type instructions with operator codes and operands. A Boolean equation, therefore, uses several rows of program, for example the equation:

$$Y = (X_1 + Y_1)X_2X_3X_7 \tag{2-2}$$

becomes (see Table 2.2):

memory number	function	type I/O	identification number I/O
0	STR	X	1
1	OR	Y	1
2	AND	X	2
3	AND	X	3
4	AND	X	7
5	OUT	Y	1

Table 2.2.

The software used for Boolean calculation must include the instruction set that can handle numerical values but gives Boolean results.

The timing function, for example, establishes a programmable timing relationship between two times which are Boolean events. This function is widely used and can be applied to manage the time parameter with good effect. Figure 2.17 shows how it is programmed using a programmable controller with Boolean equations and ladder diagrams.

(a) Pseudo-Boolean equation

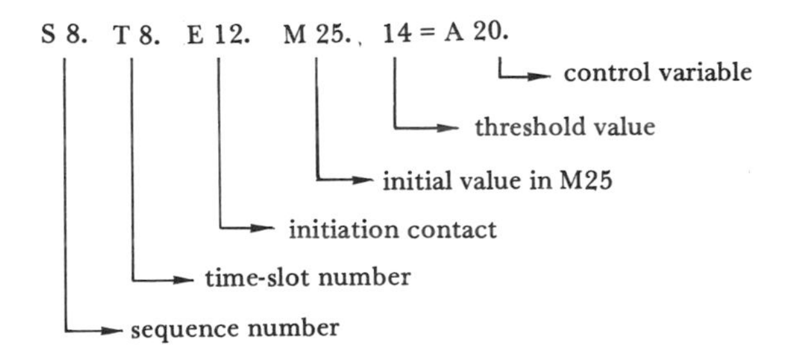

(b) Ladder diagram

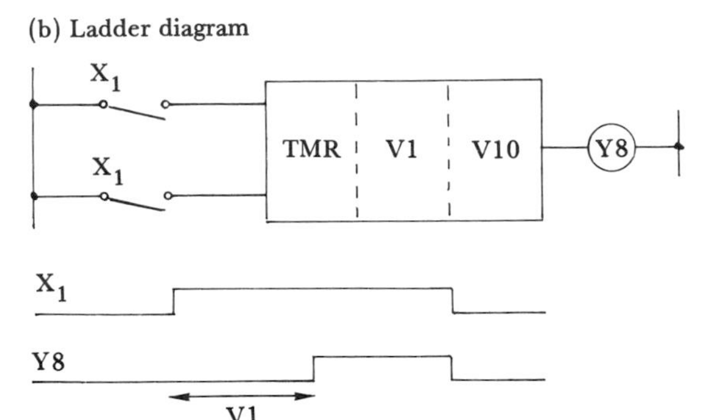

No.	Type	I/O	No.
	STR	X	1
1	STR	X	1
2	TMR	(p)	
3	V1		
4	V10		
5	OUT	Y	8

Figure 2.17. *Programming timing using a programmable controller with a Boolean equation (a) and a ladder diagram (b)*

In Figure 2.17 S 8 is the sequence number, T 8 is the time-slot number, E 12 is the initiation contact, M 25 is the initial value in M 25, 14 is the threshold value and A 20 is the control variable.

The timing function and the counting function generally arise from the same internal function. The former counts the events coming from an internal time base, whereas the latter keeps an account of the exterior events that take place in a periodic or random way.

The function used for comparison between numerical values also leads to a Boolean result and plays an essential role in industrial automatons (eg in problems of positioning, management of sequencing and weighing).

The majority of controllers are equipped with additional, more specialized functions, to augment these various standard primitives. These include:

1. a binary memory, a stepping facility and a sequencer;
2. numerical operators + − * /;
3. transcoding functions;
4. scaling function;
5. linearization by interpolation.

The basic software supplied with controllers allows them to be used in a range of industrial automatons. Section 2.7 illustrates a method for programming these devices which is independent of the basic language.

In addition to the basic methods, the experienced user will make the best use of the equipment by mastering a number of tools and techniques, such as:

1. consideration of transitions with variable logic;
2. development of a desynchronization register;
3. development of a stepping mechanism or an electronic cam;
4. generation of a horizon using the time-slot or timing function etc.

2.7 Methods for programming a programmable controller

There are many different models of programmable controller available on the market. Any programming method must, therefore, be independent of the sets of instructions for use with any of the machines. The example considered here is based on the Grafcet approach, and concerns the simple application of packing bottles into crates. This example can be used to analyse the way in which Grafcet is used to describe the control specifications of the application in a simple way. A number of methods for transcribing Grafcet onto a programmable controller are supplied. The first of these general methods makes use of Boolean equations, and is considered in greater detail before other possibilities, either currently available or still at the development stage, are discussed.

2.7.1 EXAMPLE: PACKING BOTTLES SUPPLIED BY CONVEYOR BELT INTO CRATES

The bottles arrive in a random way on the conveyor belt, and the presence of a bottle is detected by sensor p (see Figure 2.18). The pick-and-place robot stops the conveyor belt with A_1^+, and places the open gripper around the neck of the bottle with a translation T_1^+, then closes the gripper P_1^+ so as to clasp it. The bottle is thus unloaded from

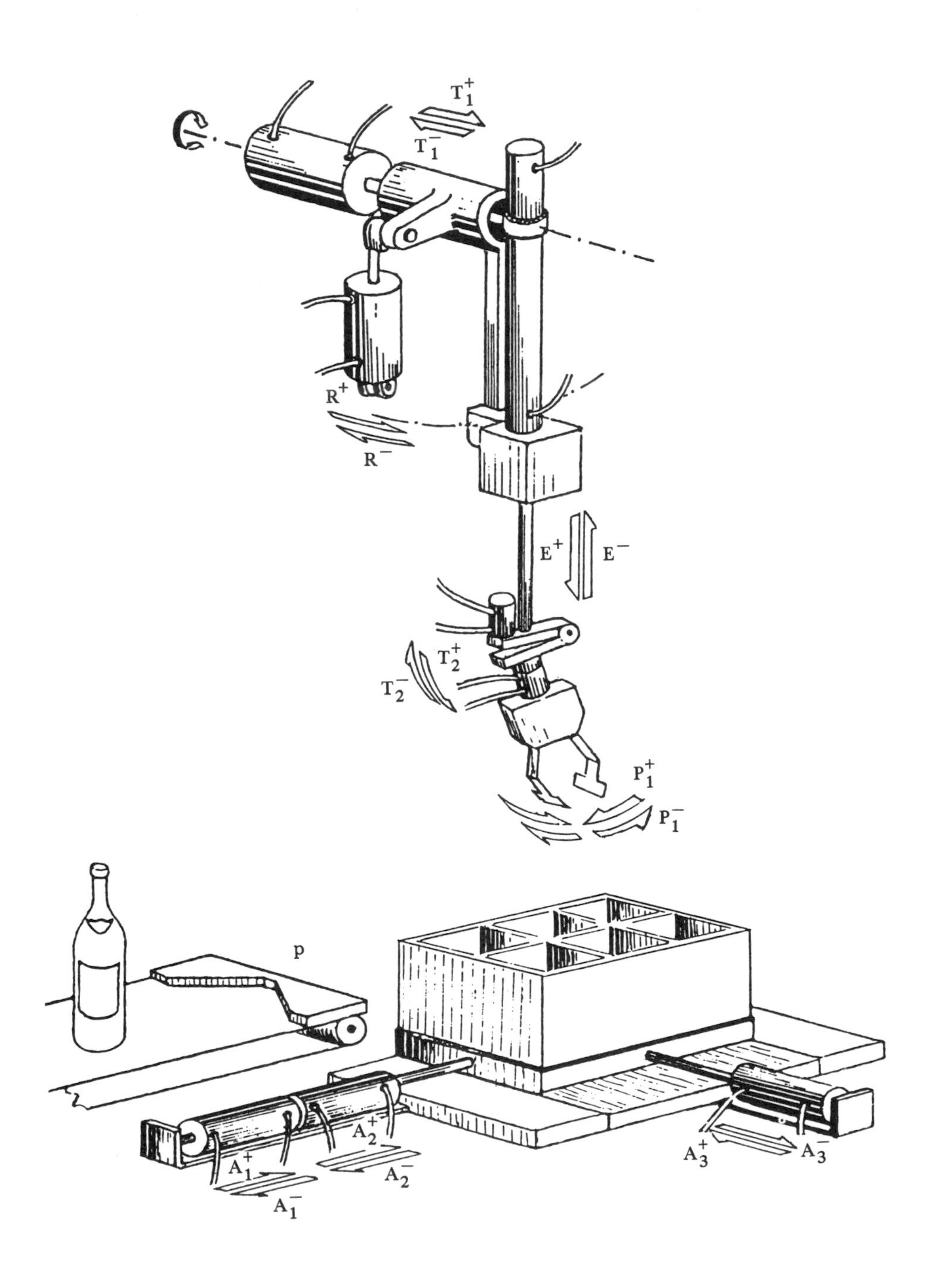

Figure 2.18. *Simplified scheme of an automated work point (bottling system)*

the conveyor belt by simultaneously effecting a torsional movement of the wrist T_2^+ and an extraction T_1^-. The bottle is raised by a return movement of the piston E_1^- and presented above the crate by a rotation R^+. The descent E^+ and depositing of the bottle are followed by the sequence of return to the initial conditions. While the arm returns, the crate is moved on the table by a set of pistons A_1, A_2, A_3, so that each of the six compartments of the crate is presented at the correct point. When the cycle has been completed, a klaxon sounds to warn the operator that he should remove the filled crate.

The inputs and outputs can be enumerated as follows:

16 inputs: $p, a_1^+, a_1^-, a_2^+, a_2^-, a_3^+, a_3^-, t_1^+, t_1^-, r^+, r^-, e^+, e^-, t_2^+, t_2^-, p_1^+$

17 outputs: $A_1^+, A_1^-, A_2^+, A_2^-, A_3^+, A_3^-, T_1^+, T_1^-, R^+, R^-, E^+, E^-,$
$T_2^+, T_2^-, P_1^+, A, M$

The actuators (outputs) are represented by capital letters, for example A_1^+ and A_1^- represent the outward and return movement of actuator A respectively. The sensors (inputs for the control unit) are represented by lower case letters, for example a_1^+ and a_1^- are the limits of the strokes associated with the outward and inward positions of actuator A.

When action A^+ is written as part of a Grafcet label, it is important to understand what is physically happening.

A^+, which can be actuated electrically or pneumatically, positions the distributor at a configuration of crossed arrows, which brings about the outward movement of piston A. This action A^+ is pulsed by nature and the distributor retains its arrangement of crossed arrows until A^- is activated to bring it back to the configuration of parallel arrows. This type of distributor constitutes an RS memory.

On the other hand, if one of the valves contains a return spring, for example for the gripper (P^- is replaced by a spring), the action cannot be pulsed and P^+ must be maintained at all stages of the gripping operation (see Figure 2.19).

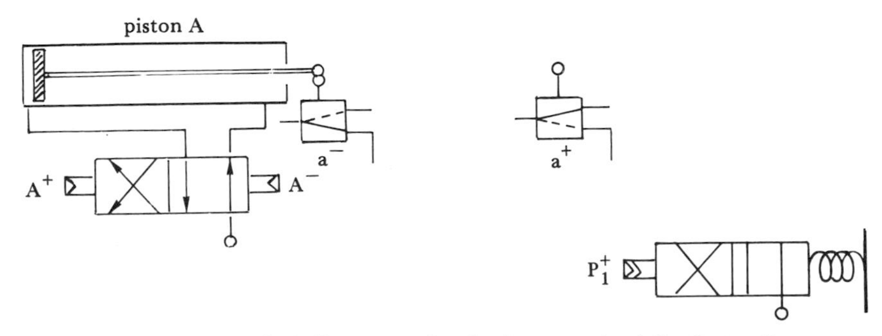

Figure 2.19. *Technical diagrams of pulsed or sustained Grafcet actions*

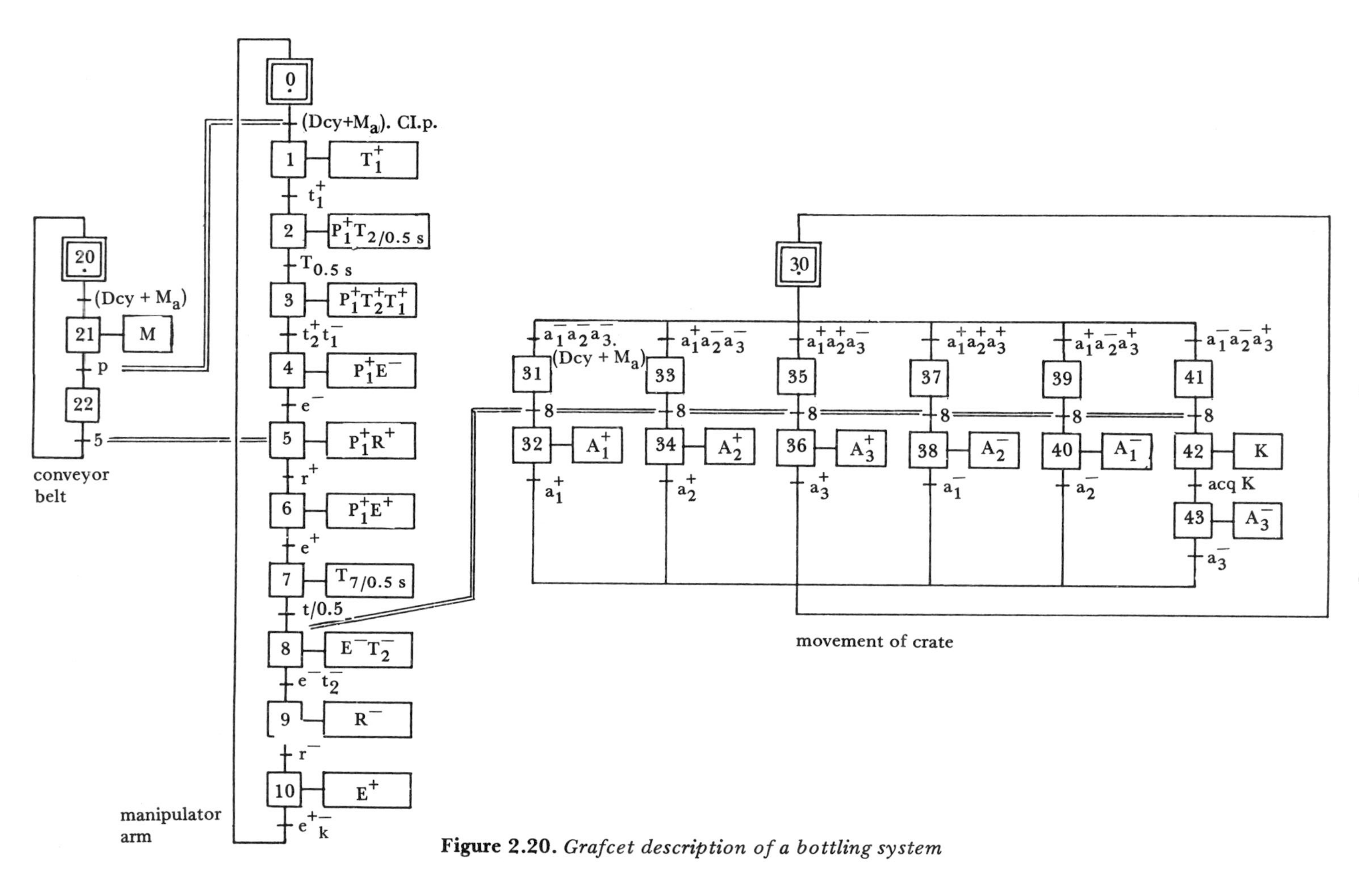

Figure 2.20. *Grafcet description of a bottling system*

Figure 2.20 shows the Grafcet description of a bottling system. The Appendix to this chapter gives definitions of the terms used in Grafcet. There are frequently several possible solutions to a given problem. Examples are given here of three Grafcets, synchronized for reasons of clarity. At the start up, the places 0, 20 and 30 are activated. D_{cy} represents the start of the cycle, and M_a the automatic operation. CI represents the initial conditions of the manipulator, thus:

$$CI = \bar{t_1}\bar{t_2}\bar{er} \tag{2-3}$$

During place 22, the conveyor belt is stopped to allow the arm to remove the bottle. By the time place 5 is activated, the bottle has been properly removed and the conveyor belt starts to move again.

As far as moving the crate is concerned, the bottles are placed in it during places 31, 33, 35, 37, 39 and 41. Place 8 of the main Grafcet, which ensures that the bottle has been released, is used as receptivity to move the crate.

The output receptivity of place 10 contains $\underline{K}$ to stop the main cycle and allows the operator time to store the full crate.

2.7.2 FROM GRAFCET TO BOOLEAN EQUATIONS: GENERAL EQUATION

At this stage, the problem the user faces is to know how to translate this Grafcet using his own controller. The 17 actions that correspond to the outputs of the programmable controller and the 16 units of sensor information will be connected with the controller inputs. The 28 places that appear in the Grafcet will be associated with the internal binary memories of the controller. From this point onwards, the Grafcet has a memory image of the controller, and the method used consists of recreating this 'graph' inside the machine. In this way, a master-slave type relationship is established between the programming device and the manipulator. Consider three consecutive places E_{i-1}, E_i and E_{i+1} separated by two transitions with which are associated the receptivities T_i and T_{i-1}. First consider place E_i (see Figure 2.21). Place E_i may be either at rest ($E_i = 0$) or activated ($E_i = 1$). For reasons of clarity, a pseudo-state: (E_i = *-activable state) is introduced.

The progression from resting state to active state cannot take place directly. It requires progression through the activable state E_i^*. Thus, when place E_{i-1} becomes active, place E_i becomes activable. Since place E_i is in an activatible state, it becomes active instantaneously once the transition T_i is reached. Place E_i, on the other hand, passes instantaneously from the active state to the resting state when E_{i+1} becomes active.

This mechanism for writing and deleting place E_i gives rise to the association of a binary memory, for example see Figure 2.22, with E_i.

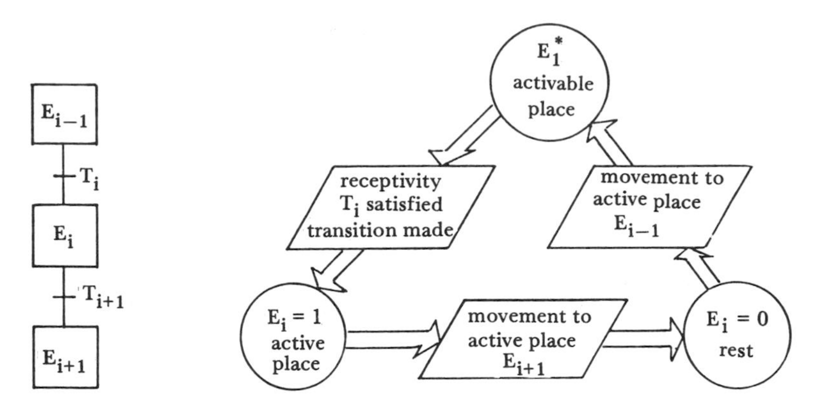

Figure 2.21. *Grafcet module and mechanism explaining changes in the state of place E_i*

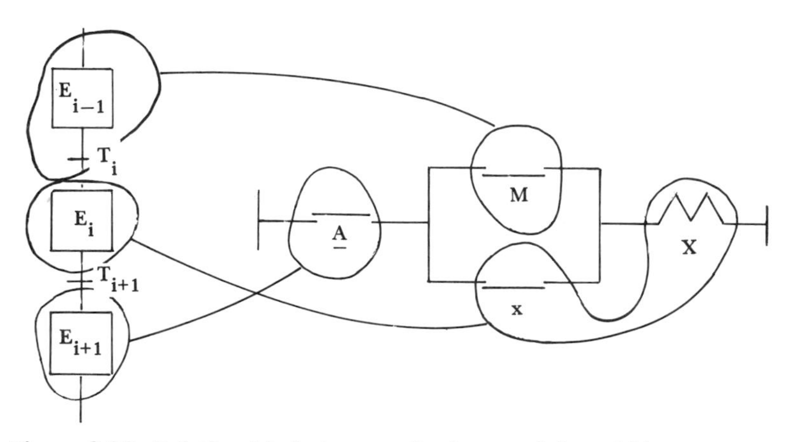

Figure 2.22. *Relationship between a Grafcet module and binary memory*

The Boolean equation for a binary memory is:

$$X = \underline{A}(M + x) \tag{2-4}$$

The connection between the automaton and the Grafcet module gives rise to the following associations:

– (X, x) the memory function with place E_i;
– (M) the set (return to 1) with an AND logic function between E_{i-1} and T_i;
– (A) the reset (return to 0) with place E_i.

The equation associated with place E_i, which is called the *general equation* since i can have any value, is:

$$E_i = \underline{E_{i+1}}(E_{i-1} \cdot T_i + E_i) \tag{2-5}$$

By writing the set of Boolean equations associated with the various places of Grafcet, and applying this process, Grafcet can be reproduced inside the controller.

The physical actions on the output level are found by attaching the action to the corresponding labels. For example if an action A_n must be carried out during phases E_i and E_j, the following can be written:

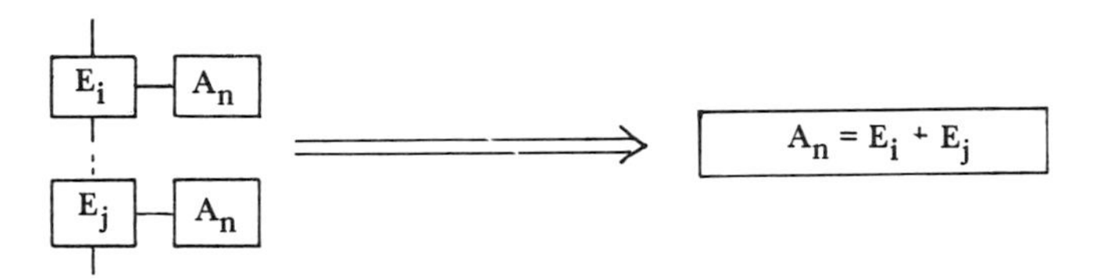

The specific characteristics of the controller, and particularly its set of instructions, are not used until the last moment to translate a set of Boolean equations. This method is thus independent of the technology.

The process used to carry out the project can be represented by the following flow chart:

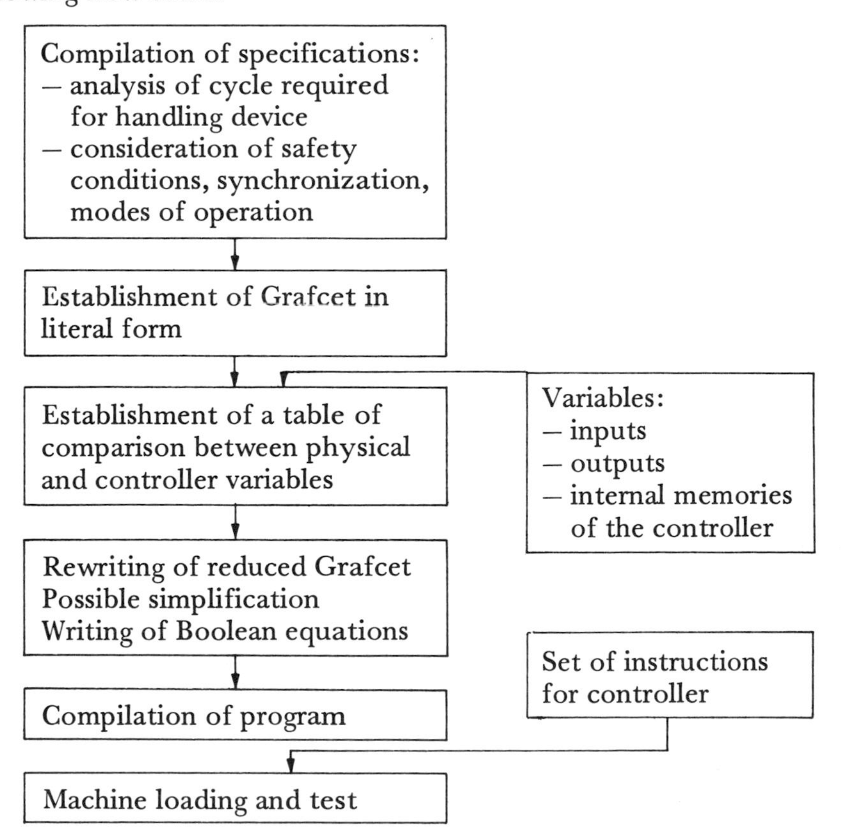

The set of Boolean equations obtained constitutes a universal representation and thus a language (Pascal p-code or Iso language in numerical control). Its translation into the set of instructions for a particular machine requires an interpreter function, which is provided by the programmer himself. A certain number of controllers with Boolean notation (eg the Renault SMC) are particularly well suited to this approach.

Note 1: The associated equations can be written in two consecutive places:

$$E_i = \underline{E_{i+1}}(E_{i-1} \cdot T_i + E_i) \qquad (2\text{-}6)$$

$$E_i = \underline{E_{i+2}}(E_i \cdot T_{i+1} + E_{i+1}) \qquad (2\text{-}7)$$

As soon as the receptivity associated with T_{i+1} is satisfactory, E_i is deactivated and E_{i+1} is activated instantaneously and simultaneously. Considering the sequential and cyclical operation of controllers, one of the equations will be executed before the other, and the two places E_i and E_{i+1} will both be active during a controller cycle, ie for a duration of a few milliseconds.

The two physical actions A_i and A_{i+1} are activated simultaneously over a period of a few milliseconds. The time may vary in different situations, but it is usually much shorter than the response time of electromechanical actuators. A safety precaution can be taken by establishing that:

$$A_i = E_i \cdot \underline{E_{i+1}} \qquad (2\text{-}8)$$

or by reversing the order of the equations E_i, E_{i+1} in the program. This type of detailed analysis is not valid unless the details of the mechanism that take into account the inputs and applications of the outputs for the controller used are known (see Figure 2.23).

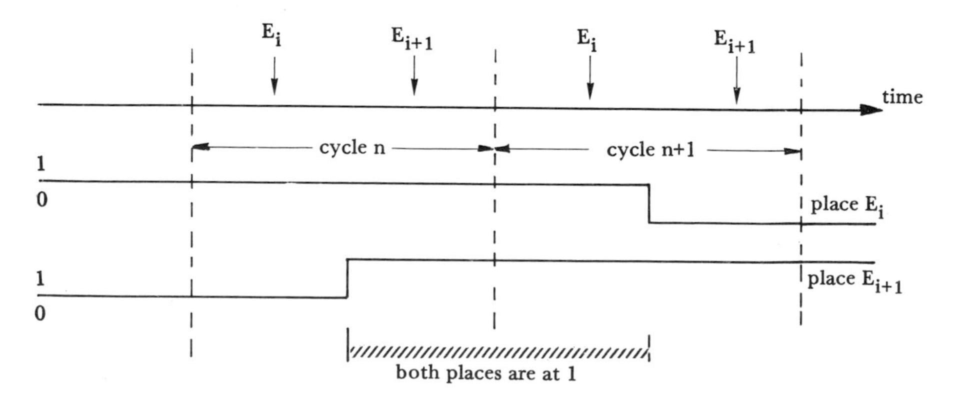

Figure 2.23. *Mechanism capable of generating two consecutive simultaneously active places*

Note 2: Grafcet rules specify that if, during operation, a single place must be activated and deactivated simultaneously, it remains activated. The general Boolean equation adopted does not satisfy this rule since priority is given to deactivation. If this situation arises in an application, the Boolean equation associated with a priority engaging memory can be adopted:

$$E_i = E_{i-1} \cdot T_i + \underline{E_{i+1}} \cdot E_i \qquad (2\text{-}9)$$

Note 3: Taking into account the AND or OR divergences or convergences of the subprograms, loops and backwards or forwards jumps involves the application of a special but straightforward equation (Laurgeau *et al.*, 1979). In particular, the development of subprograms for switching, and loops etc using relays may be considered a direct consequence of this method.

It would be impractical to present the Boolean system for the example of the crates and bottles because of its length, but it is possible to establish a simplified Grafcet which clearly displays the method used. The set of Boolean equations that corresponds to this Grafcet is shown in Table 2.3.

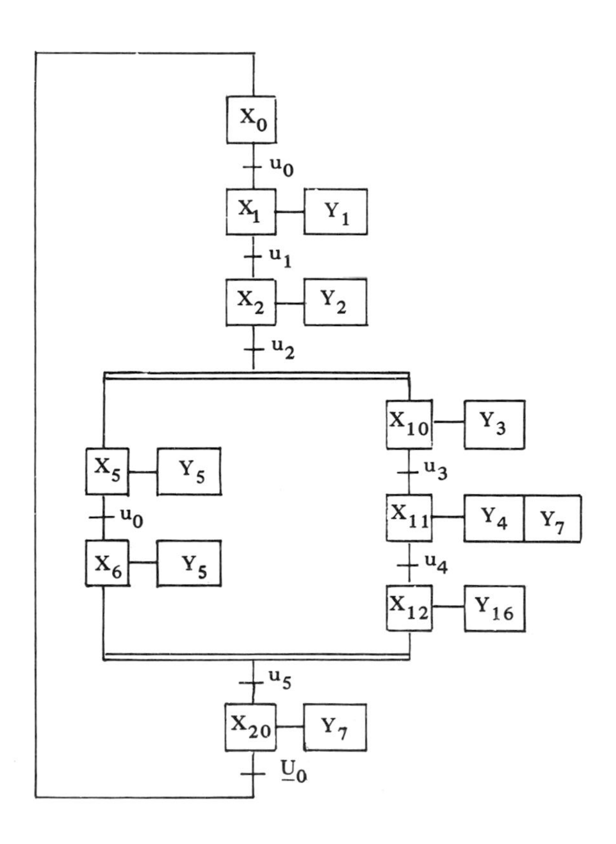

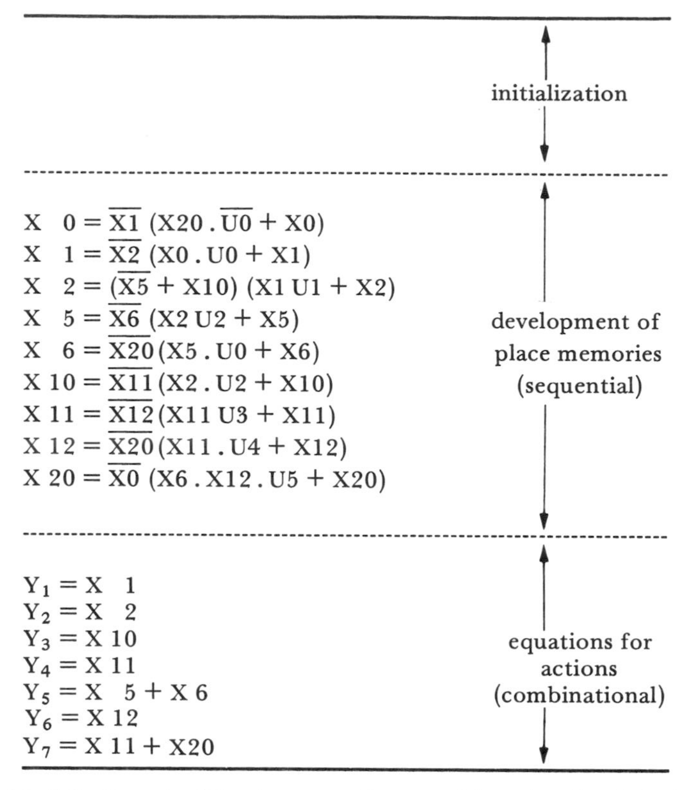

initialization

$$X\,0 = \overline{X1}\,(X20\,.\,\overline{U0} + X0)$$
$$X\,1 = \overline{X2}\,(X0\,.\,U0 + X1)$$
$$X\,2 = (\overline{X5} + X10)\,(X1\,U1 + X2)$$
$$X\,5 = \overline{X6}\,(X2\,U2 + X5)$$
$$X\,6 = \overline{X20}\,(X5\,.\,U0 + X6)$$
$$X\,10 = \overline{X11}\,(X2\,.\,U2 + X10)$$
$$X\,11 = \overline{X12}\,(X11\,U3 + X11)$$
$$X\,12 = \overline{X20}\,(X11\,.\,U4 + X12)$$
$$X\,20 = \overline{X0}\,(X6\,.\,X12\,.\,U5 + X20)$$

development of place memories (sequential)

$$Y_1 = X\,1$$
$$Y_2 = X\,2$$
$$Y_3 = X\,10$$
$$Y_4 = X\,11$$
$$Y_5 = X\,5 + X\,6$$
$$Y_6 = X\,12$$
$$Y_7 = X\,11 + X20$$

equations for actions (combinational)

Table 2.3. *The set of Boolean equations used in a simplified Grafcet*

2.7.3 OTHER METHODS FOR TRANSCRIBING GRAFCET ONTO A PROGRAMMABLE CONTROLLER

There are several other methods of carrying out this transfer. Because of the success achieved by Grafcet users, the manufacturers produce specific methods adapted to their own products. The instructions provided by the manufacturers contain solutions for programming Grafcet using their machines. Consider the following examples.

2.7.3.1 The stepping method

Some controllers include the stepping function, familiar to all technicians, in their set of instructions.

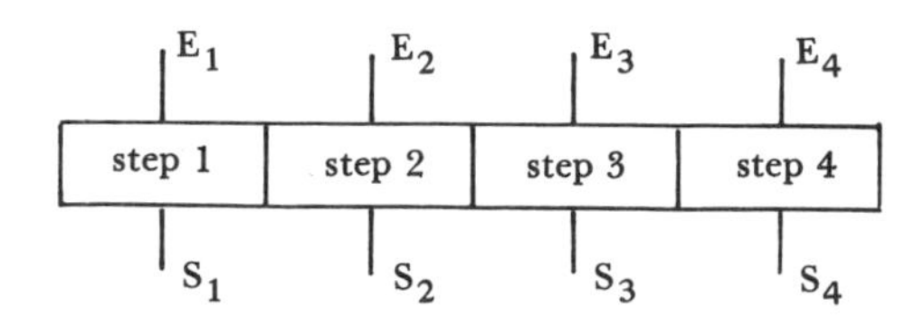

Step 2 is activated if step 1 is activated, and if the condition E_2 is at the logic state 1. The activation of step 2 automatically brings about the deactivation of step 1 and the actuation of output 2. This stepping process also involves the sequencers or the materialization of linear Grafcet, and provides a solution which is both powerful and neat.

Some controllers have a facility for graphics stepping. The stepping process can be displayed on a VDU by simply defining the:

1. conditions associated with the inputs E_1, E_2 . . . (receptivity);
2. actions associated with the outputs S_1, S_2 . . . .

The main significance of this type of representation is the:

1. graphics display of dynamic states allows the user to see the cycle develop;
2. structure is already created so few instructions are required to put it into practice.

Receptivity is frequently limited to one variable; when this is the case, it is necessary to transcribe a more complex receptivity in an internal variable. On the other hand, the linear and fixed structure of the stepping operation is not sufficiently flexible to show divergences, convergences, jumps and place repeats.

2.7.3.2 Method for conditional evolution

The development of a place E_{n-1} towards place E_n is conditioned by tests.

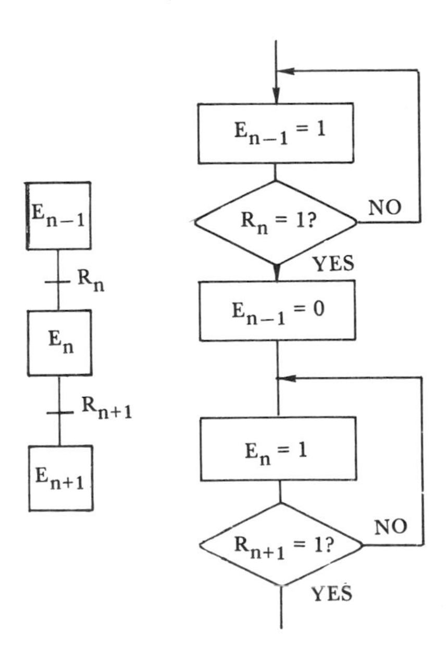

A type of program writing will be established for the controllers in which the instructions 'IF . . . THEN' will appear:

IF 'TEST' THEN 'ACTION'

Many controllers which do not have this instruction have an alternative in the form of a downwards jump. The jump will avoid the execution of the actions:

TEST { place E_{n-1} active and receptivity R verified

JUMP if NOT TEST

ACTION { -activation of E_n -deactivation of E_{n-1}

2.7.3.3 Direct Grafcet language

This language is not yet commercially available from the manufacturers, but it is the object of current research, and uses graphics techniques, including colour, to program the controller directly. This method of graphics programming is already used in conjunction with ladder diagrams, particularly for American programmable controllers. The use of graphics would allow the specifications to be seen directly in the form of places, transitions etc onto the VDU. The possibilities for display are limited by the size of the VDU in the same way as ladder diagrams. It is, however, conceivable that the whole VDU console could be considered as a window moving in all directions on an electronic page, with the possibility of a magnifying glass effect. The dynamic development of Grafcet could be displayed using tokens, a flashing effect or colours. This type of VDU console associated with a controller is a product with a general application and would be in no way specifically linked to robotics. Along the same lines, it is possible to imagine controllers adapted to a specific type of manipulation in which programming is carried out in graphics form, not with Grafcet but directly onto an animated image of the robot.

2.8 Implicit programming by training

2.8.1 MECHANICAL PROGRAMMING

Purely mechanical automatic devices have been in use for many years. Consider, for example, weaving looms, agricultural machines, sewing machines and transfer arms. These ingenious devices contributed to the progress of the industrial revolution, and they are still used today in many production workshops. Control of trajectories and speed, and the development of sequences of cyclical actions are all carried out by the automaton itself, by an *a priori* choice of the appropriate architecture and mechanisms. A large number of devices can be used in constructing an automated mechanical system: screw-nut systems, roller cams, cylindrical or conical drum cases, wheels with pawls, mechanical regulators, drums with picks, punched cards etc (see Figures 2.24 and 2.25).

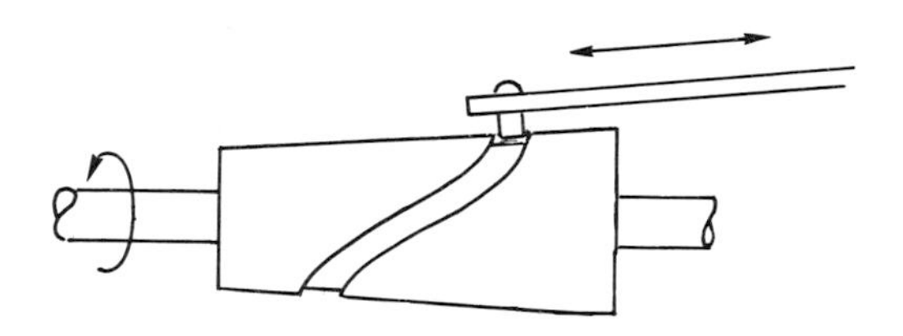

Figure 2.24. *Example of a cam with conical roller used to obtain translation based on complex speed laws*

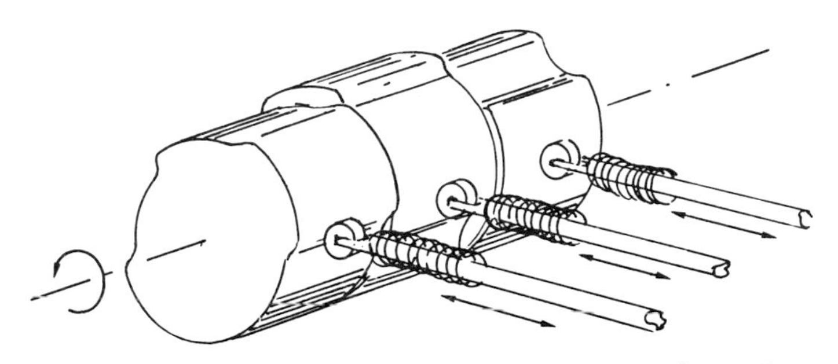

Figure 2.25. *Drum with rollers used to form complex cycles*

Consider the way in which a complex trajectory in a plane can be programmed using a single motor. Points A and B are in contact with two different cams causing translational movements along y and x at speeds y(t) and x(t) determined by the cams. The trajectory of the end effector f(x, y) results from the combination of speeds x and y. A single motor can control the axes x and y (see Figure 2.26).

It may seem excessive in this context to use the expression 'mechanical programming', but the communication between man and machine necessary for the user to control the machine efficiently does exist at this stage of machine design. There is no use of language or training, for example, but when programming is carried out at a much later stage in the design process, the automaton is fixed in its specific type of action.

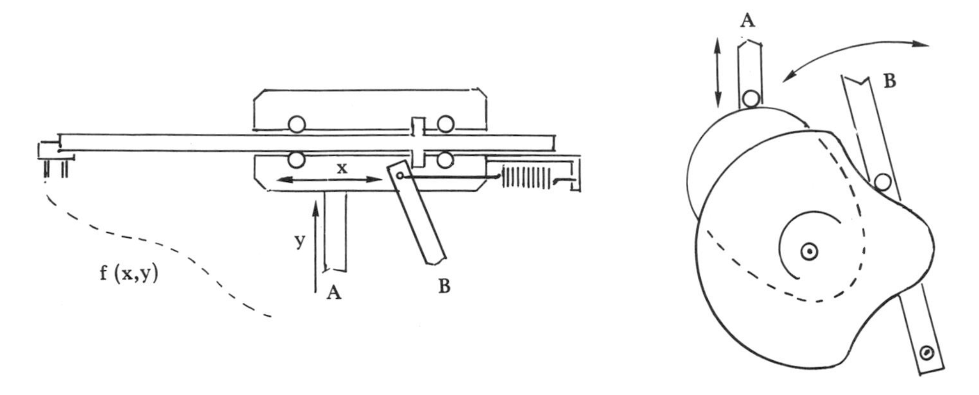

Figure 2.26. *Principle of a mechanically programmed two-dimensional trajectory*

2.8.2 DIODE MATRICES

In the majority of pick-and-place robots, the amplitude of the linear or rotary movements in the various axes is controlled partly by mechanical stopping points which are moved manually and bolted into place. The adjustment operation requires the manipulator to be stopped. Some robots are equipped with manually adjustable stops that can be electrically actuated during the cycle. This solution constitutes a progression from the one mentioned above, and two or three programmable stopping points are accessible on certain axes (see Figure 2.27).

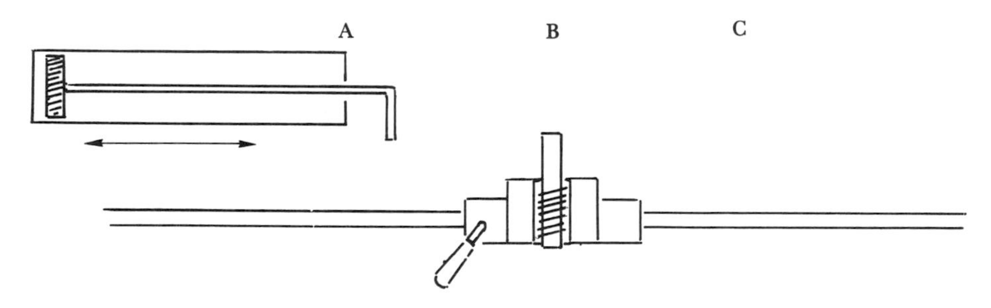

Figure 2.27. *A and B are fixed and C can be put into use electrically*

A more elaborate system (eg the Versatran or the Shinko robot) uses a diode matrix with a potentiometer matrix. The limits of the strokes are regulated using potentiometers, which involves positional servo-control of each axis. The diode matrix contains as many rows as there are axes controlled and the number of columns determines the maximum number of stopping points in the cycle. The potentiometer matrix contains as many rows as there are servoed axes and the number of columns determines the number of possible stopping points on each axis. At each point of the training cycle, the axes that are moved are positionally adjusted by the potentiometers. Several axes may be activated in the course of a single step, giving rise to trajectories not parallel to the axes, but in the absence of speed control, subsequent steps are activated when the slowest axis reaches its final position.

Programming with diode matrices on fixed mechanically adjustable or potentiometer-regulated stopping points is straightforward and does not require any particular expertise. Nevertheless, this technology which is used in many materials handling robots is being incorporated less and less into newly developed lines.

2.8.3 TRAINING SYSTEMS

When programming it is important to consider the problems both from the point of view of the user and the designer. The majority of playback robots in use in industry are programmed by training. High-level languages are becoming increasingly more popular for programming assembly robots.

The application of a method of programming by training or by language can create an awkward problem for the designer and yet be quite straightforward for the user. For the user, programming a robot by teaching it the task to be performed can be carried out in minimum time and requires no specialized training in computing.

In the same way, the user of pick-and-place robots may find the idea that he will have to program the mechanism himself, either on a programmable controller or a microcomputer, daunting. For this reason, manufacturers attempt to provide pick-and-place robots with the same training facilities as playback robots. There are relatively few systems that can be used in a technically simple manner. The same is true of specialized languages adapted for the end effectors of pick-and-place robots. However, it is likely that in years to come these products will be better developed (see Table 2.4).

	pick-and-place robots	playback robots
programming by training	– programming matrices – Duplico-type training system	– direct training by syntaxer, by telecontrol, or according to point-to-point or continuous path trajectories
programming by language	– direct programming on programmable controller – specialized language for sequence automatons	– symbolic language on actuator, end effector, object or objective levels

Table 2.4. *The differences between programming by training and programming by language for pick-and-place and playback robots*

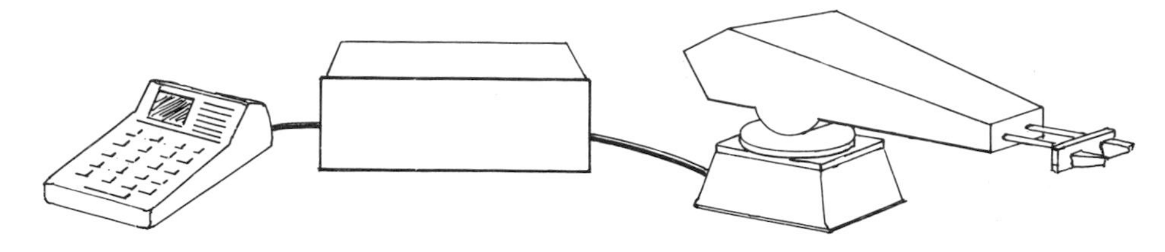

Figure 2.28. *Training box-programmable controller-manipulator unit*

To illustrate this point, consider the example of Duplico. As already described, programming a programmable controller connected to a manipulator is straightforward but involves problem analysis. The programming is avoided if a Duplico training box is used (see Figure 2.28). The training box removes the need to program, in that when it is coupled with the programmable controller it can make the manipulator carry out the required cycle by a direct command from the user. When the required cycle is achieved the training box is disconnected and the work program of the manipulator is recorded. The cycle can then be repeated automatically and indefinitely. The characteristics of the training box allow the following functions to be used:

1. adaptation to any handling device (with up to five axes) and many programmable controllers;
2. registration of movements on limit switches or timing;
3. operating modes:
   - axis by axis, manual;
   - training;
   - continuous path or stepping verification with correction functions;
   - storage/loading of programs;
   - running;
4. control of ancillary actions external to the manipulator;
5. synchronization with external sensors on the manipulator;
6. possibility of sequential branching with external synchronization;
7. detection of sensor faults.

Figure 2.29 shows the training console and manipulator and Figure 2.30 shows the keyboard in detail.

The originality of this solution and the method adopted by the designer lies in that it is oriented towards the final user to a high degree. Considered in the context of robotics, the training box has the same relationship with the programmable controller as the controller itself had with the process computers in the context of production computerization. The success of programmable controllers is due in part to their potential as problem-oriented language machines. The standard universal computer language, which is complex to handle, is replaced

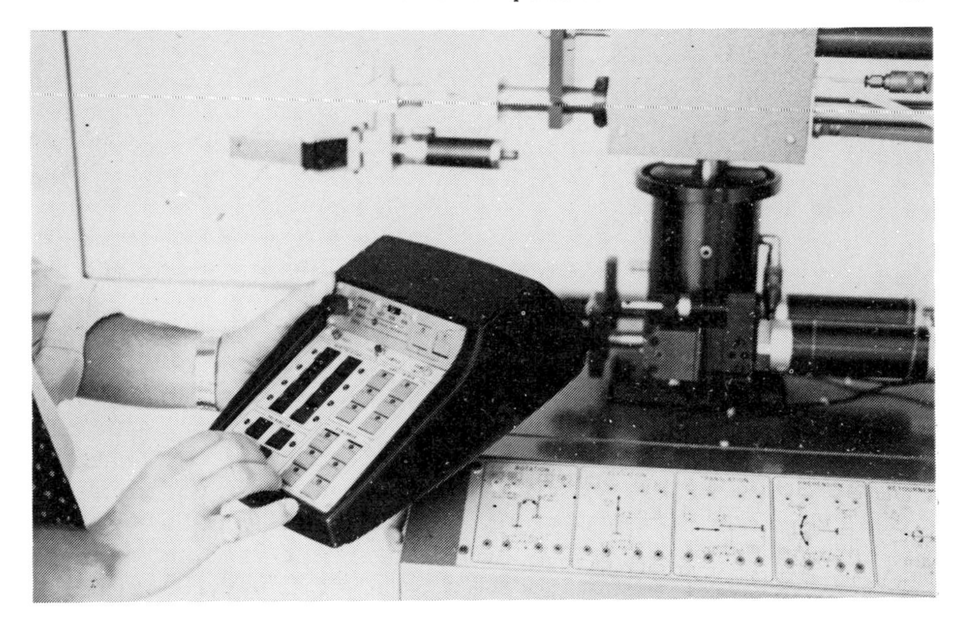

Figure 2.29. *Duplico training keyboard (from Valoris)*

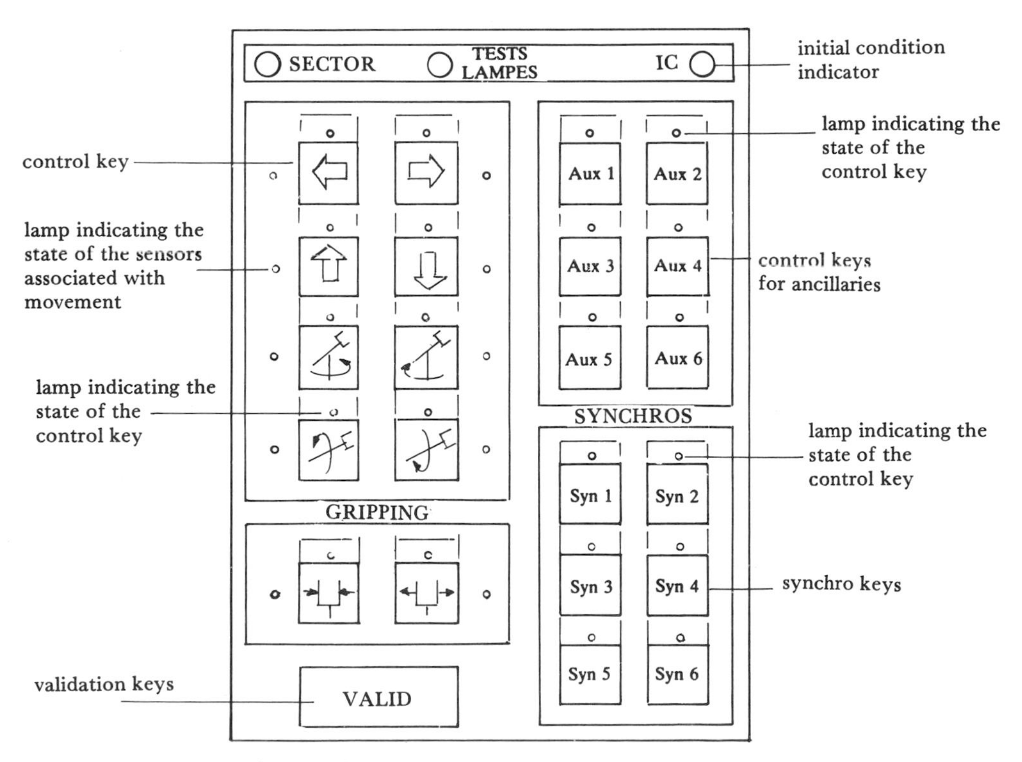

Figure 2.30. *Details of the training box keyboard*

in the programmable controller by a reduced set of primitives oriented towards production. This controller language is replaced by a smaller number of primitives relating to manipulation of the type: extend arm, move to left, raise, grip etc. This solution is an indication of the developments to be expected in the future. It should be noted, however, that the training box is doubly dependent on the manipulator for its source language, and on the controller for its translated language. The training box, when used to replace a syntaxer or keyboard, acts as a translator, from an end effector level language for a maximal effort manipulator into a controller language. The training box can be substituted for the programming console of a controller, but in context will lead to a less than complete use of the controller's potential.

The programmable controller and its console make up a processing tool independent of the processes to be automated. The solution given above is not the ultimate answer, and will, no doubt, be replaced in the future by a control board integrated into the manipulator, which will be less expensive and directly programmable by the training box.

Appendix

Concepts and rules for the development of Grafcet

Main concepts

Grafcet is a tool used to describe the control specification of an automated system. It involves a number of concepts.

1. *Place*, symbolically represented by a numbered square, corresponds to a situation in which the behaviour of the automated system, either partly or as a whole, does not vary with regard to inputs and outputs. The possible actions associated with a place are described inside a rectangular label, to the right of the place. These actions are not effective unless the place is activated.
2. *Transition*, symbolically represented as a barrier separating two consecutive places, shows the relationship between places. Associated with each transition is a logic condition called *receptivity*, which, if fulfilled, allows the transition to be fired.
3. *Connections or arcs* are oriented connections linking a place with a transition, or a transition with a place, but never linking two places or two transitions together.

Grafcet is made up of all sets of places, arcs, transitions, receptivities and labels forming a representative graph of the operation of an automaton.

Rules

RULE 1

The initialization specifies the places activated at the start of operation. They are activated unconditionally. The initial places are shown on Grafcet, by doubling the lines of the square symbolizing the place.

RULE 2

A transition is either validated or non-validated. It is validated when all the places immediately preceding are active. It cannot be made unless it is validated, and unless the receptivity associated with the transition is true. If these conditions are satisfied the transition must be made.

RULE 3

If a transition is carried out, it simultaneously and instantaneously brings about the activation of all the immediately following places and the deactivation of the immediately preceding places.

RULE 4

Several transitions may be carried out simultaneously, if this is intended in the design.

RULE 5

If, in the course of operation, a single place must be simultaneously activated and deactivated, it remains activated.

Chapter 3

Servocontrolled robots

3.1 Introduction

In Chapter 2 it was shown that simple robots based on modular units can be adapted to carry out a variety of tasks by mechanically adjusting their movements and reprogramming their cycles of operation. Such robots are generally used in simple operations in which the flexibility of the unit is restricted. For example, in loading and unloading a press or machine tool, the robot would be assembled specifically for use with this machine, and its flexibility would be limited to the range of parts to be handled often by mechanically changing the gripping device. A robot of this type cannot usually be used for a different task or an identical task on a machine with a different architecture. Simple robots usually offer little flexibility, but as they generally have the advantage of being modular in design they can be assembled easily and at low cost from standard components. This approach, which has been in use since the earliest days of automation, is particularly suitable for long production cycles in which the robot is used for simple tasks with little variation.

Despite the simplicity of the constituent parts, each development of a new robot-based application can be expensive and time-consuming. It is generally advisable to entrust the pilot study to a person with a specialized knowledge of robotics. The more complicated points of the study, such as those of breaking the task down into steps and the choice of parts as a result of this analysis, development of software (hard-wired or by controller) and adjustment of the finished product, can take several months. The total time required to develop and install a typical system for assembly (see Figure 3.1) may be of the order of one year.

The hourly cost (C_h) for a unit of production equipment (automated or not) can depend on many factors:

C_{eq}: total cost of parts;
α_{eq}: residual value (expressed as a percentage) of the parts that can be reused after the application is no longer required;
C_e: cost of design and development;

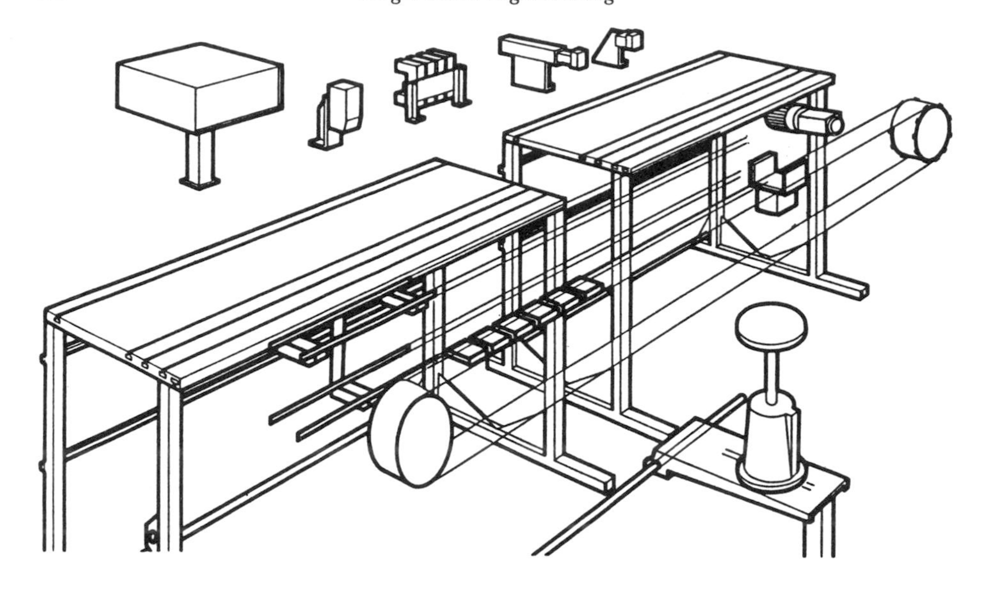

Figure 3.1. *Sormel assembly module*

T_c: average length of time between two production modifications;
C_c: average cost of each modification;
C_f: cost of the operation per time unit;
T_t: total length of operation expected of unit (minus periods when not in use).

The equation representing the hourly cost is:

$$C_h = \frac{(1 - \alpha_{eq})(C_{eq}) + C_e}{T_t} + \frac{C_c}{T_c} + C_f \qquad \textbf{(3-1)}$$

This simple formula can be used to compare very different solutions to a production problem, by assuming the other parameters to be equal or negligible (ie product quality, rapidity of response to market changes, availability of manpower). It can be used, in particular, to compare an unautomated process with both an automated process with low levels of flexibility and a flexible automated process.

Programmable robots are generally used for work requiring flexible automation. They can be justified on the basis of:

1. marked increase in the cost of reusable parts and also in the intrinsic value of the machine;
2. reduction in costs of study and adaptation;
3. reduction in the average cost of modifications to the product.

The increase in costs of parts and of salaries cannot fail to reinforce the advantages of using more flexible systems instead of manual or fixed sequence manipulators. Despite the problems in establishing precise economic justification, the advantages shown below are fundamental in encouraging car manufacturers to automate their production lines using robots to whatever degree of complexity they choose.

parameter / production	C_{eq}	α_{eq}	C_e	T_c	C_c	C_f	T_t
manual	very low	very low	average	poor	very low	high	poor
fixed automation	average	average	high	high	high	average	high
flexible automation							
now	high	high	average	any	very low	average	any
in the future	↘	↗	↗	↘			↘

Table 3.1. *The advantages of flexible automation now and in the future*

3.2 Servocontrolled industrial robots

Unlike simple manipulators in which the ends of the strokes can only be adjusted mechanically, servocontrolled robots can be programmed over the entire length of the strokes of each of their programmable axes, that is they can reach any point in their work space that has been set by training or calculated during any single program.

In practice, a robot can only take up a limited number of positions on each of its degrees of freedom (DOF) because of the control technology used, which will be considered in detail later, but in comparison with the techniques already considered the range of positions that the robot can reach is considered continuous.

A robot may have mixed mobility, that is, have some degrees of freedom subject to servocontrol and others (in particular for certain movements of the wrist or gripper) with bang-bang control. Even if only one DOF is flexible, however, the robot is classified as servocontrolled. Volkswagen uses modular welding robots in its factories, some of which are simplified versions with only one DOF, and yet they are classified as robots. The system which allows each flexible axis to take up a sequence of positions for a given application can be broken down as shown in Figure 3.2:

1. *Memory*: this is the fundamental part of the programmable robot. The information stored in the memory enables the robot to carry out tasks which may vary in complexity but which are in essence variable. Robot programming implies storing information for the

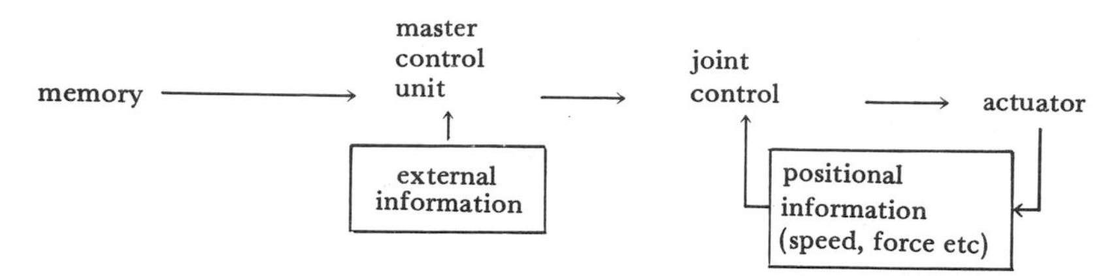

Figure 3.2. *Actuator control*

required task. Robot memories are now almost universally digital. Their implementation may be by integrated circuits or magnetic tape (eg disks, tapes, bubble memories).

2. *Master control unit*: this unit is used to create the commands required for control from the information stored in the memory, and possibly also using external signals. The master control unit may, for example, be required to calculate the many intermediate points in a trajectory (whether rectilinear or curvilinear) defined by two or three points stored in the memory.
3. *Joint control unit*: this is the interface between the processor (which stores and processes data) and the actuator. It can consist of an electrical servo-system, a hydraulic servovalve and its associated circuitry, a control unit for stepping motors etc. Depending on the method used (servo-system) the joint control unit can utilize a feedback from the actuator.
4. *Actuator*: as the name indicates, this is the mechanical part of the robot, which is generally made up of a motor, step-down gear and transmission system. In a servocontrolled robot, the actuator will cause the joint being controlled to reach a large number of positions. These various elements are described in detail in this chapter. The order in which they are considered is arranged logically in terms of comprehension of the programming method. The first element considered is the actuator in association with its electrical control role in the four basic processes: electrical stepping, electrical servocontrol, hydraulic servocontrol and pneumatic servocontrol. Then, the memory is described followed by the master control unit.

The chapter concludes with a description of the programming methods most commonly used in the three major types of servo-controlled robots programmable by training:

1. point-to-point control robots;
2. calculated trajectory robots;
3. recorded trajectory robots.

Chapter 4 describes programming using languages and not training. Chapter 5 describes programming from a CAD system.

3.3. Actuators and actuator control

In the industrial robots currently in use, the actuator and its control when activated by a low-voltage electrical signal from the master control unit cause a movement of the terminal device or end effector, or at least of the axis under consideration. It seems likely that in the near future, some robots, in addition to controlling their movements, will also be able to control the degree of force used, for example in assembly or machining operations. In robots currently in use, the only control is over position, and the force is supplied by ancillary compliance devices, which operate with varying degrees of efficiency.

The discussion here is limited to a description of how an actuator can take up a certain position according to an electrical signal and considers force control as a possibility only. Since industrial robots are made up of mechanical parts of non-negligible mass (even if they weigh only a few grams), in every case a source of power external to the master control unit is required. Depending on the nature of the source and the way in which it is used, controlling the actuator will vary. Without giving a detailed account of the current technology, which is discussed in detail in Volume 4 of this series (Lhote *et al.*, 1984), the main modes of control of industrial robots are now considered with regard to the parts used and the constraints imposed on the master control unit.

3.3.1 ELECTRICAL ACTUATORS

Electric motors intended for use in robotics must satisfy a certain number of conditions. They must:

1. be able to cope with changes in speed, and must be able to maintain a position when bearing a load and at zero speed;
2. possess total reversibility in direction and function;
3. be able to produce high maximum torque, particularly at low speeds;
4. possess a low mechanical time constant T_m to avoid reducing dynamic performance;
5. possess a high thermal time constant T_{th} to reduce temperature increases which can alter performance;
6. be subject to little oscillation in torque as a function of the rotor position;
7. possess a linear speed-torque relationship.

3.3.1.1 Direct current motors

Eighty per cent of all continuous path control robots use this type of motor. The inductor flux is generally constant and is provided by permanent magnets (eg Alnico, ferrite, rare earths). The motor is controlled by varying the armature supply. These motors are frequently referred to as torque-motors, because of the low speeds (< 1,500 rpm) at which they function and the linearity of the speed-torque relationship. Since the inductor flux is constant, the electromagnetic torque is proportional to the current flowing through the armature. Torque is therefore assessed by measuring current, and the torque can be controlled by supplying the motor using a current generator (see Figure 3.3).

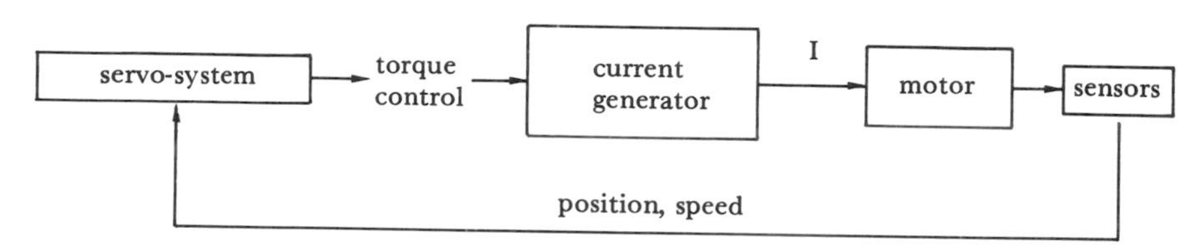

Figure 3.3. *Servo-system loop*

There are many versions of the direct current motor available. Only two types are discussed here: the axial motor and the flat torque-motor.

The *electric motor* is the most widely used, and it is used particularly for powering numerically controlled machine tools. Many manufacturers produce these motors with a torque range of between 0.5 and 120 Nm (see Figure 3.4).

The *flat torque-motor*, frequently known as the *pancake* motor, is characterized by a large diameter-to-length ratio (often greater than 5 : 1). Its architecture and the fact that it is available in kit form (eg armature, step-down gears, brush system) allow it to be integrated directly into

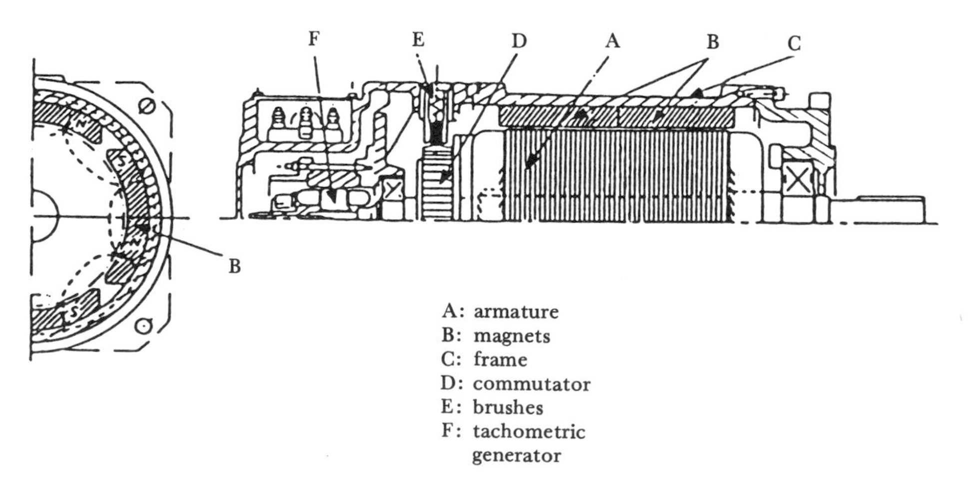

Figure 3.4. *Electric motor*

Figure 3.5. *Artus flat torque-motor*

the joints, associated with a positional encoder and a harmonic drive type flat step-down gear, or even connected directly (see Figure 3.5).

This is particularly useful in wrist joints with two or three axes, swivels and gripping devices etc. The most common armature control voltage is 70 volts but it can reach 150 volts in large systems. From the point of view of interfacing with the robot control processing unit, there are two methods available, depending on whether the axis servo-system is autonomous and under analog control (see Figure 3.6) or whether the servo-system is under digital control (see Figure 3.7). Despite the fact that numerical regulators with microprocessors have been in use for several years in slow industrial processes (eg petrochemical, paper, food industries) the problems of numerical control in rapid electromechanical processes have not been fully resolved.

Developments in software and technology are, however, pointing the designers towards the production of cards or the numerical control of axes using specialized microprocessors with microprogrammed multiplication and division operators (type 6809 or 8087). It should be noted that if precision of 1/10 mm for an axis with a stroke of 1 m is required, 10,000 points must be used. The precision of the converters and the calculation words should not be less than 14 bits. This suggests that 16-bit microprocessors are the best suited for controlling robot systems.

3.3.1.2 Stepping motors

Unlike direct current motors in which the angular position integrates the control voltage of the armature, the angular positioning of stepping motors is achieved by discrete successive increments in response to

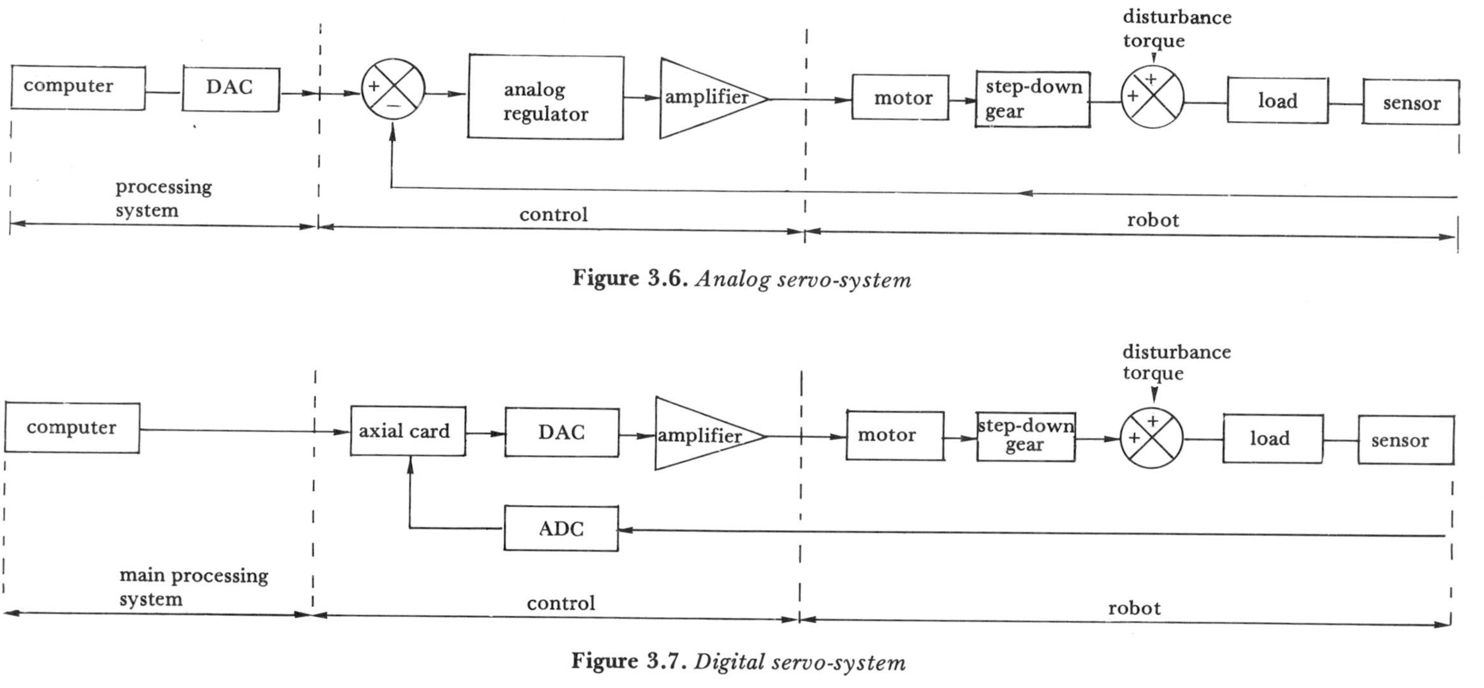

Figure 3.6. *Analog servo-system*

Figure 3.7. *Digital servo-system*

Figure 3.8. *Stepping control*

control pulses. The number of steps per revolution varies between 200 and 1,000. Therefore, stepping motors have two main advantages:

1. positional control by counting-deducting pulses without the use of a positional sensor;
2. speed control by careful application of pulse frequencies without use of a tachometric generator (see Figure 3.8).

These advantages give rise to an open-loop mode of operation, but are not realistic unless the unit is reliable and does not miss steps. The limitations to this attractive system result from the maximum power levels obtained (1.5 kW), the complexity of the electronics required for control and the impossibility to control torque. It should be noted that some manufacturers of programmable controllers include primitives for the management of stepping motors in their basic instruction set and specialized chips have also been developed.

3.3.2 HYDRAULIC ACTUATORS

Hydraulic energy is nearly always used to power heavy robotic operations. Its most significant advantages are:

1. a high power-to-mass ratio (more than 1 kW/kg for a pressure of 100 bars) which allows motorization to be distributed without making the terminal devices of the robot too heavy, and avoids using transmission links;
2. the time constants are very low and the dynamic performance very good in respect of speed and acceleration.

There are, however, also disadvantages:

1. high cost is a result of the need for the basic architecture (hydraulic central unit) and of the price of some of the components (eg servovalves);
2. hydraulic central units are noisy, although actuators are silent;
3. there is a risk of leaks;
4. it is difficult to control the force generated by the actuators.

The actuators themselves can be classified according to the linear or rotary motions generated. For linear movements, single- or double-action pistons, with single or double shafts are used. For rotary movements, a large variety of hydraulically powered motors are available, with gears, flappers, radial or axial pistons etc. Neither the pistons nor the motors are controlled directly. They are controlled via other components whose function is to decide the direction of motion and its amplitude. These components are distributors and servovalves.

3.3.2.1 Distributors or electrovalves

These distributors are used to control the flow of the fluid and to direct it in one direction or another. The spool, which constitutes the major part, moves linearly and can take three stable balance positions: position I which allows a given direction of motion, position III which allows motion in the opposite direction and position II which prevents motion (see Figures 3.9 and 3.10). Position I is achieved by activating electromagnet E_1, position III is achieved by activating electromagnet E_2 and position II is naturally stable and obtained without activation. The information required for control, which is provided by the processing system, is in this situation a Boolean variable either pulsed (⎍) or sustained (⎍) depending on the electrovalves used.

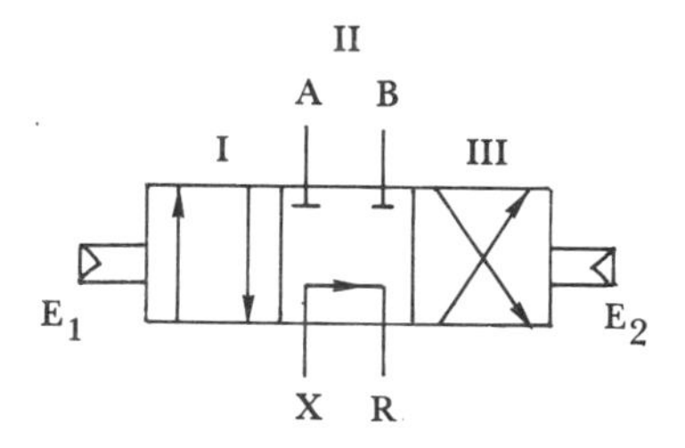

Figure 3.9. *Controlling a motor using electrovalves*

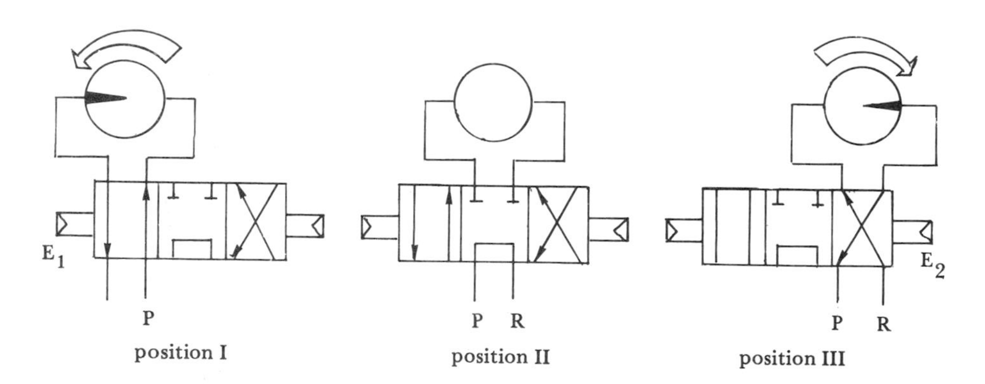

Figure 3.10. *Controlling a motor using electrovalves*

Bang-bang control may sometimes be adequate for a large number of applications and provides positional control. If, for example, the linear speed of axial movement is 1 m/s and the commutation time of the electrodistributor is 1 ms, the positional precision would be of the order of a millimetre, which is acceptable for many applications.

3.3.2.2 Servovalves

Like electrodistributors, servovalves control the flow of the fluid, and can direct it accordingly, but in addition they can modulate the outgoing flow by between 0 and 100 per cent of its nominal value. This provides a more precise control over position and makes direct speed control of the actuator possible. The electrical control power of servovalves can vary between a few milliwatts and a few watts, but power amplification is considerable (of the order of 10^5 and can be as high as 10^9). This amplification is carried out in one or more stages. The first stage of the servovalve may be made up of a spool, a nozzle-flapper or jet deviation, but the later stages always take the form of a spool. Seen from the outside, control of the spool, flapper or pipe is maintained using a torque-motor or a proportional electromagnet. The processing machine must deliver a control current between 0 and 200 mA through a converter and suitable assembly (see Figures 3.11 and 3.12).

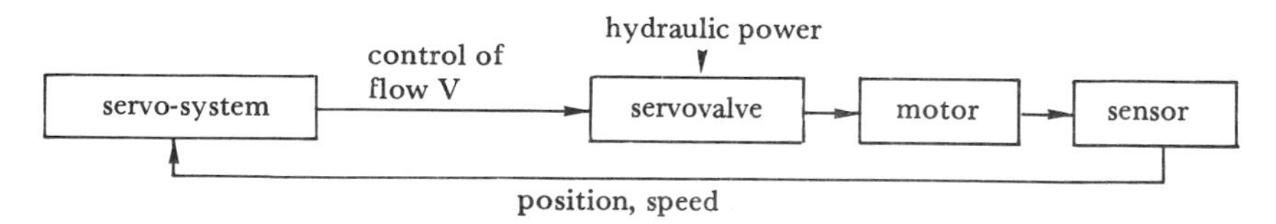

Figure 3.11. *Scheme showing control by servovalve*

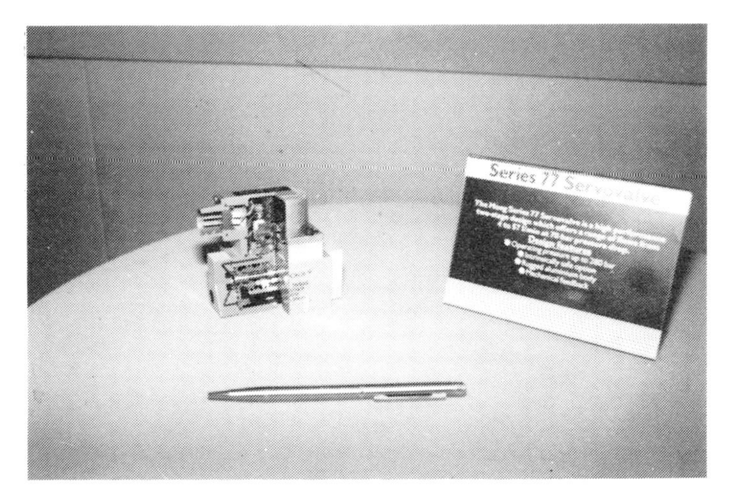

Figure 3.12. *Moog servovalve*

3.3.3 PNEUMATIC ACTUATORS

Pneumatic actuators (see Figure 3.13) are very similar to their hydraulic equivalents. The basic difference results from the use of air,

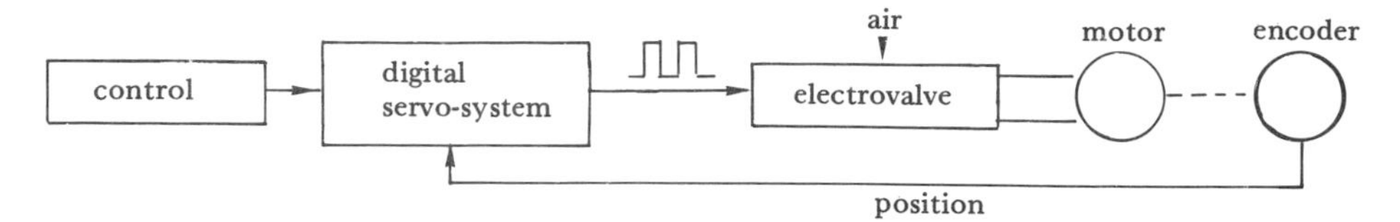

Figure 3.13. *Pneumatic servo-system*

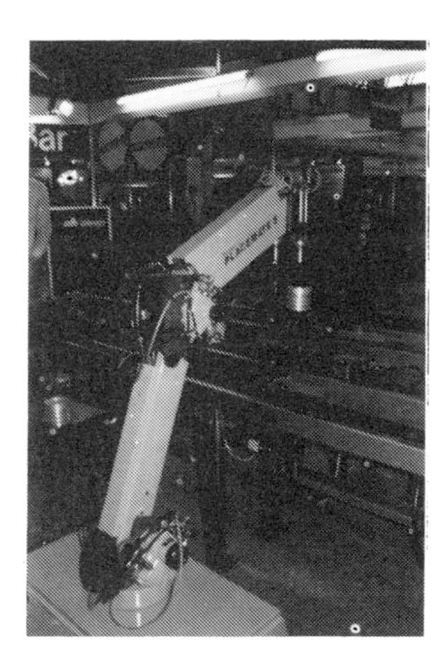

Figure 3.14. *Pandar pneumatic robot*

a compressible gas, instead of an incompressible fluid. The compressibility of air ensures that the position of the moving part is dependent on load, and it is therefore difficult to impose positional servo-control with variable loads. This explains why pneumatically driven actuators are usually only used in the pick-and-place robots considered in Chapter 2. Some American and British manufacturers are beginning to produce servocontrolled robots driven by pneumatic stepping motors, which offer a good performance-to-price ratio (see Figure 3.14). The advantages of using pneumatically driven systems are:

1. low cost;
2. very high power-to-mass ratio (up to 100 W/kg for a pressure of 10 bars);
3. easy transport and storage.

On the other hand, this type of mechanism is noisy because of the escape of air from the open valves. It is sensitive to pollution and above all does not easily allow position, speed and force to be controlled. From the point of view of the control unit, the control delivers Boolean variables which activate, in most cases, electromagnetic on-off valves.

3.4 Memory

To carry out a program involving its various actuators, an industrial robot must store data on the task in its memory. Such data, which since the days of the first prototype have been stored numerically, can be:

1. positional data which represent the action of an actuator or the final position of a tool, or the position of an object in a particular set of coordinate axes;
2. algorithms which allow successive positions to be calculated, such as in the case of palletization;
3. authorization which must specify timing for the operations in progress coordinated by external signals.

In all situations, the master control unit, which is described later, is responsible for the program required to initiate the robot motion as a function of these three types of information.

3.4.1 POSITIONAL DATA

In most industrial robots, positional data directly represent the movements of the actuator. The actuator stroke (linear or circular) is limited to a value C, which is generally less than the mechanical stroke C′. This software-generated stroke C (as opposed to the mechanical stroke C′) is divided into a certain number of positions P. P is related, as shown previously, to the type of actuator and its control (eg stepping motor, digital servocontrol) or it can simply be dictated by the constraints of using binary memories.

Each position in stroke C is represented by a binary number. Since, in general, the binary numbers are expressed in words with a fixed length of n bits,* the following relationship can be written:

$$P \leqslant 2^n \qquad (3\text{-}2)$$

Since computers use 8-, 16- or 32-bit words, the value of n is usually 16 (or two 8-bit words), which leads to a maximum of:

$$2^{16} = 65{,}536 \quad (64\,\text{K})$$

points in the stroke.

* In data processing a *word* is a series of n binary values (0 or 1). These binary values (or bits) can thus express 2^n combinations (from 000 . . . 0 to 111 . . . 1). Numerical values can be attached to these combinations by stating that each word is the binary representation of a number. In this way 2^n values (from 0 to 2^{n-1}) are found. It is also possible to consider the first bit as a sign bit, thus allowing negative and positive values to be expressed:

$$-2^{n-1} \text{ to } 2^{n-1}-1$$

Other systems of representation allow numbers to be expressed on a larger scale of values but with a smaller degree of precision (floating point numbers).

In general, P is less than 2^n, since this allows the placing of the points defining a software-generated stroke C, strictly contained within the mechanical stroke C′. It is sometimes difficult for technical reasons (see Section 3.3) to make the end points required for the actuator correspond to the end point positional values (eg 00 . . . 0 and 11 . . . 1). In certain robots, moreover, these end points vary with the positions of the other actuators. The robot designer will therefore place software-generated stopping points (possibly variable) on the software-generated stroke. Between the two software-generated end points will be the P discretized values of the stroke.

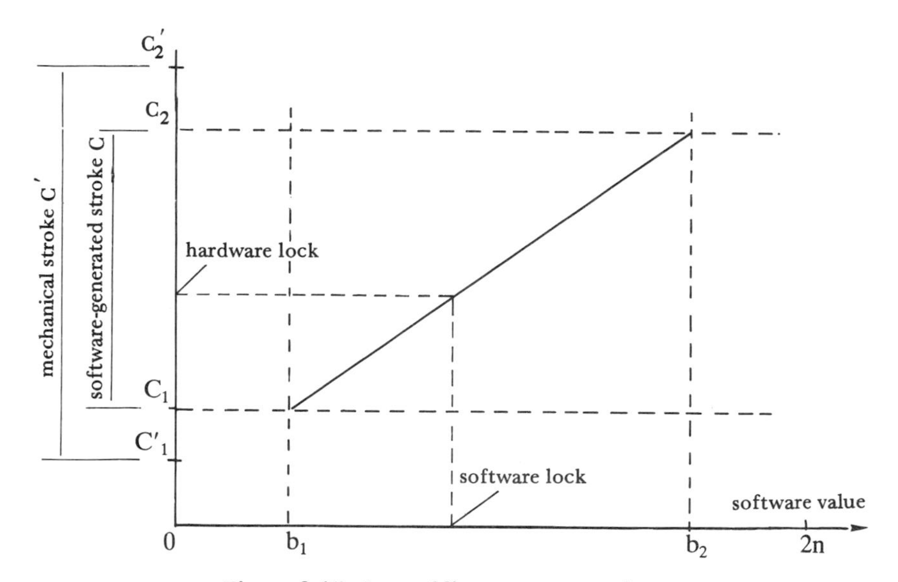

Figure 3.15. *Law of linear correspondence*

The relationship between a software-generated value of the stroke and a real value (see Figure 3.15) is created by choosing:

1. a locking position, that is a software-generated value corresponding to a precise mechanical value which can be located in a set of coordinate axes by the control part (eg an electrical stop or the top of an optical encoder). On the axes which do not use an absolute encoder (eg stepping motors, two-track optical encoder) this locking operation must be carried out each time the robot is used;
2. a law of correspondence between a binary value and a mechanical value for the stroke. This law is linear, but trigonometric laws could also apply, and would allow, for example, the calculations for control in a Cartesian work space in the case of rotary articulations to be simplified.

For each software-generated value there is a corresponding theoretical value for each actuator and if a law of linear correspondence is assumed to exist, the logic resolution (R_1) of this DOF is given by its software-generated stroke divided by P:

$$R_1 = \frac{C_2 - C_1}{P} \tag{3-3}$$

Taking an extreme, but real, example, with a stroke of 8 m and a word of 16 bits in which all the values are used to represent the stroke, the following logic resolution is obtained:

$$R_1 = \frac{8\,\text{m}}{2^{16}} = \frac{8}{65{,}536} = 0.122\,\text{mm} \tag{3-4}$$

This is an extreme case because of the very long stroke. The logic resolution is generally good and markedly better than the physical resolution and precision resulting from the control parts (eg resolution of a digital-analog converter, optical encoder, stepping motor, potentiometer linearity, servocontrol error). It is also often better than the mechanical precision of the actuator (eg backlash, non-linearity, imprecision of dimensions). However, the logic resolution is often presented by manufacturers as the physical resolution (and sometimes as precision).

Using digital representation as a basis, it may appear easy to memorize a succession of robot positions by using a word of n bits per DOF for each configuration. When the programs are sequential, instead of storing the absolute value of the position for each servo-system for every point of the trajectory, it is equally as effective to store only the relative value of this position in relation to the preceding point. From a starting position, such as the locking position, it is easy for the master control unit to update each absolute position for the various DOF by using the relative values supplied at each step. This method allows the number of bits necessary for storing the information at every point in the program to be reduced. On the other hand, this method reduces the maximum interval between two consecutive points in a program.

The method of relative recording is of interest when the number of bits to be stored is large, either because of the high value of n (eg axis with a very long stroke) or because of the many points (eg robot with a continuous trajectory). There are, nonetheless, three problems which arise in storing a series of points:

1. method of classification;
2. storage space required;
3. time required for access to a new point.

These three problems are directly related to the different technology available for storing information and the application in mind.

	capacity (8 bits)	random access time	flow of 8-bit words per second	can stored information be modified?	can stored information be safeguarded?	detachable input
core memories						
magnetic core	tens of thousands	$< 1\ \mu s$	$> 1{,}000{,}000$	yes	yes	no
RAM	tens of thousands	$< 1\ \mu s$	$> 1{,}000{,}000$	yes	several hours with batteries	no
PROM	tens of thousands	$< 1\ \mu s$	$> 1{,}000{,}000$	no	yes	possible
EPROM	tens of thousands	$< 1\ \mu s$	$> 1{,}000{,}000$	off-line	yes	possible
EEPROM	tens of thousands	$< 1\ \mu s$	$> 1{,}000{,}000$	yes	yes	no
bubble memories	several hundreds of thousands	1 ms	100,000	yes	yes	possible
disks						
Winchester	several tens of millions	50 ms	100,000	yes	yes	no
floppy 8 inch	several hundreds of thousands	100 ms	50,000	yes	yes	yes
floppy 5 inch	several hundreds of thousands	100 ms	50,000	yes	yes	yes
magnetic tapes	several hundreds of millions	several seconds or minutes	1,000,000	yes	yes	yes
punched tapes	several tens of thousands	not possible	100	no	yes	yes

Table 3.2. *Characteristics of core memories*

Central memories

These are memories which can be addressed directly by the control device (central processing unit). There are many types available which are characterized by their rapid *access times* (of the order of microseconds) and high costs (and so reduced capacity: typically tens of thousands of 8-bit words).

Magnetic core memory: it will probably disappear from the market because it cannot compete with the cost of semiconductor memories. It is independent of interruptions in power supply and so can retain information indefinitely.

RAM (Random Access Memory): it is a semiconductor memory which holds information that can be modified or lost. Hence, it requires emergency batteries to safeguard the stored information if the power supply is disconnected (which is adequate for a few hours or a few days for RAM CMOS).

PROM (Programmable Read Only Memory): it is a semiconductor memory which permanently stores information that cannot be modified by the processor. REPROM can be reprogrammed on a special unit after deleting the stored information with ultraviolet light.

EEPROM (Electrically Erasable Programmable Read Only Memory): the information in this memory is stored permanently, but the memory can be reprogrammed by the processor. If the power supply is interrupted the information is not lost.

Bubble memories

These are frequently considered as extensions of the core memory because they usually cannot be detached, and so the user may be unaware of their separate existence. They are well adapted to industry as they include no moving parts, can store information permanently, even in the event of a loss of power, and can be modified by the processor. The access time is of the order of milliseconds and these memories cost less than semiconductor memories but more than disk or magnetic tape memories. The capacity is of the order of a few hundred thousand 8-bit words.

Magnetic disks

When first produced, these required careful handling but they are now used increasingly more in industry. They are available in different diameters (and so with different storage capacities), and can be divided into two types: *hard disks*, which remain integrated in the machine and *floppy disks* which allow programs to be stored outside the machine. The capacities range from a few hundred thousand to several million 8-bit words, with an average word access time of the order of 1/10 s, but with significant throughput in sequential use.

Magnetic tapes and punched tapes

These tapes are characterized by the fact that the information they contain can only be read sequentially, and can only be used for initial loading of a program into the core memory. Even on this level, they are rivalled by floppy disks. It seems likely that only mini- and microcassettes in the simplest systems will survive.

To conclude, depending on the application, these criteria will impose very different constraints on the organization and technology of the memory. If all robots must store data corresponding to the program to be carried out in memory, it must be remembered that data can vary widely with the type of trajectory. In Sections 3.4.2 to 3.4.5 the different types of trajectory are explained with respect to memory.

3.4.2 POINT-TO-POINT TRAJECTORIES

In this method, only a few characteristic points of a trajectory are recorded by the robot, along with the conditions for moving from one point to another (eg an external event and/or a signal sent to the exterior required to initiate an action). This type of robot is most commonly used for machine loading and unloading and spot welding. The only important parts of the trajectory are the destination positions and possibly some intermediate points which allow obstacles to be avoided. These intermediate points, although not imposing the same constraints as the destination points, are treated in the same way from the point of view of storage in memory. At the moment of task execution, the robot moves from one point to another, giving each axis its destination position, which must be reached in the shortest possible time. There is no coordination between the axes, and the trajectory is often unpredictable and sometimes not constant in length (see Figure 3.16).

The program of a point-to-point control robot will, therefore, be a logical series of assorted positions with parameters affecting the conditions for motion towards the next point. Generally speaking this will take the form of a Boolean equation relating to a set of binary inputs, but more usually the true values of a single signal. Also included are the binary ouputs to be activated when the correct point is reached

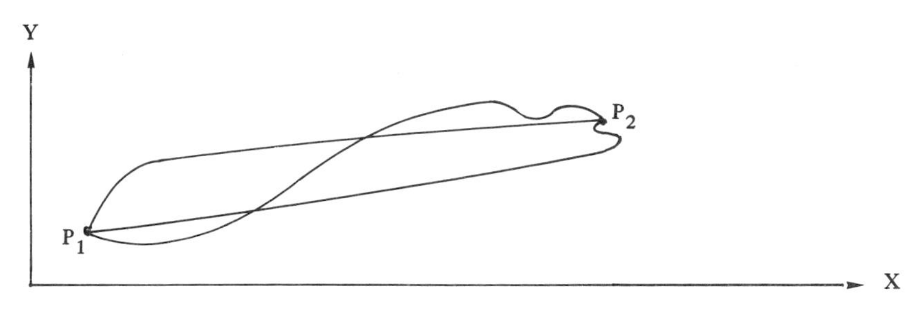

Figure 3.16. *Examples of a point-to-point trajectory in a plane*

(these may, in some cases, control some bang-bang axes in the robot and its end effector). The timing, which must be observed before moving from one point to another, is also provided by the program. This feature is frequently used to stabilize the robot in its final position when the servo-system does not inform the master control unit that the point has been reached and/or that the robot is subject to significant vibrations.

No. of point	type
position of axis 1	
position of axis 2	
. . .	
position of axis n	
conditions of chaining	
output variables	
timing	

Table 3.3. *Variables relating to one point in the program controlling a point-to-point control robot*

The variables relating to one point of the program are stored in a data block. The length of the data block depends on the characteristics of the robot (eg number of axes, logic resolution, complexity of input-output operations) but will never need to exceed a few tens of 16-bit words.

The number of points in the program is never large (a few hundred at most), and for this reason memory capacity is not large either, except in cases where the robot requires access to a large number of programs, as may be the case for welding robots on a production line where a number of different models, for example of car, are produced.

The maximum time required to assess a new point is of the same order of magnitude as the time required for the robot to move from one point to another, that is about a second, which does not restrict the choice of input system. Since the progress of a program is always the same, the blocks can be stored sequentially in the system.

Despite these favourable conditions which appear to support using the simplest type of information storage (eg the first type of memory used in the Unimate system was magnetic drums, but even punched tapes could be used), the small quantity of information required for a program is now stored in the central memory of the master control unit at the moment of execution. This method of storage in the central memory has the advantage that it does not require sequential use of blocks which are linked to one another by pointers, that is data representing the addresses of the preceding and following block in the memory (see Figure 3.17). This method makes editing programs easier, that is it allows points in the middle of the sequence to be deleted or inserted without the need to copy out the new program physically. These modifications are very easily made by modifying the pointers themselves, and the new blocks can be created in any available space in the memory. During execution of the program, the master control unit is required simply to follow the list of pointers n + 1 in order to find the programmed sequence.

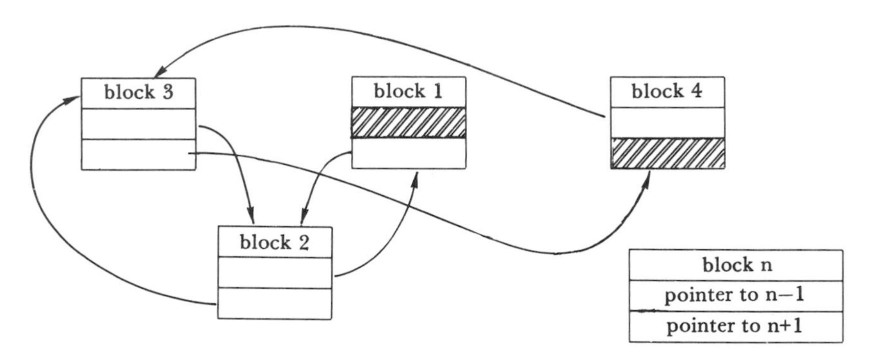

Figure 3.17. *Chaining blocks in a program*

3.4.3 CALCULATED TRAJECTORIES

Calculated trajectory control robots are similar to the ones described above with the exception that the trajectory generated between two points is constrained by the programmer. The programmer specifies the geometry of the trajectory and its kinematic constraints.

This type of robot is most commonly used for arc welding, where one of the key parameters is the speed with which the torch moves.

Although this type of robot is distinctly more complex than the previous example, when considering the actuators (ie the stability and precision of the motors and servo-systems must be better) or the master control unit (which is described later) the information stored in the memory is similar. It is, of course, necessary to add to each block data on the type of trajectory, which can be:

1. non-constrained;
2. linear in the articulation space;
3. linear in the Cartesian space;
4. circular;

and on the speed of movement of the tool (specifying whether the intermediate points are stopping points or passing points). This supplementary information is compactly stored (it can be contained in a single 16-bit word), and the programs for these robots reach a maximum of a few hundred points, as before, and so constraints relating to the memory space available are not restrictive and the programs can easily be stored in the central memory of the master control unit. This was shown to be the case with point-to-point control robots, and the advantages are those discussed previously in relation to programming.

3.4.4 RECORDED TRAJECTORIES

This type of robot is most widely used for painting. Unlike the previous example, it is not possible to generate the trajectory from little information. The motions involved are complex, since they must copy those of a human operator, whose movements whilst painting do not follow any simple kinematic law.

In theory, only an analog recording of the required trajectory will enable it to be reproduced accurately. In practice, the precision (or repeatability) imposed will result in high costs, and may even be technically impossible. Since the most modern techniques of analog recording involve digital conversion it is easy to understand why it is used by continuous path robots. The trajectory is sampled at times $t_1, t_2, t_3 \ldots$ when the values of the DOF are collected and stored in memory. The sampling period is constant, and it is fixed by either the master control unit (with a high level of precision, using a quartz clock, or more approximately, using a variable clock if it will be necessary to act upon the speed of execution in a continuous manner) or a periodic external command. The last method is used in particular when robot movement must be synchronized with the displacement of the part. In this situation, the conveyor supplies the sampling command to the recording unit, then to the restitution mechanism which allows the robot motions to be accelerated or decelerated according to the speed of conveyance (this adjustment is slight, since it is difficult to relate the flow of paint to the speed of the robot).

Since performance standards of the paint-spraying robot are high (speed is 2 m/s and acceleration is 1 g for the AKR 3000 robot), the frequency of sampling must also be high, so that maximal performance can be maintained. If a trajectory with a speed of 1 m/s is sampled at millimetre intervals for the length of the stroke, the sampling should be 1 ms, or a frequency of 1,000 Hz. In practice, sampling frequencies of less than 100 Hz are adequate since rapidly generated trajectories are usually linear because of the mechanical constraints on both the arm of the human operator carrying out the training and on the mechanical arm.

Despite this, the sampling frequency supplies much information. Consider, for example, a robot with six axes, each of which supplies information in 12-bit words (this would result in a resolution of a millimetre for the standard paint-spraying robot). The flow of information obtained at 100 Hz is:

$$100 \times 6 \times 12 = 7{,}200 \text{ bits/s}$$

or 900 bytes/s. This means that a 32 K central memory represents only 35 s of program, and a 500 K diskette less than ten minutes.

For this reason, manufacturers of continuous path robots have adopted techniques for condensing data. There are two main methods available:

1. *Linearization by segment.* This method allows the frequency of recording for each of the axes to be reduced to recording simply the values and times which would allow reconstruction by segments, whilst minimizing a particular criterion (eg smallest square root or limited error) (see Figure 3.18).
2. *Binary code.* Unlike the preceding method, this technique retains the same recording step as in sampling, but reduces the quantity of information stored at each point. Instead of storing the new

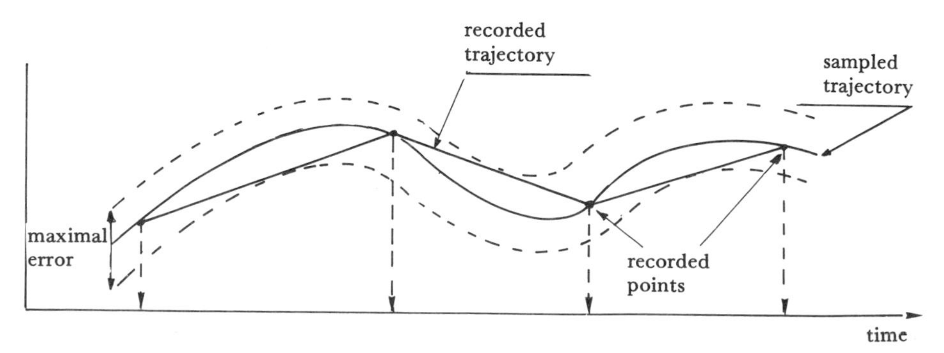

Figure 3.18. *Linear modelling with limited error*

information as an absolute value, it is simply recorded (with an adequate degree of precision) in terms of its relationship to the preceding point. It is possible to record, in binary code, the position of the last point calculated relative to the sampling curve (below or above). A refinement to this method is to record the acceleration instead of the speed, also in binary values. The starting point must, of course, always be recorded as an absolute value (see Figure 3.19).

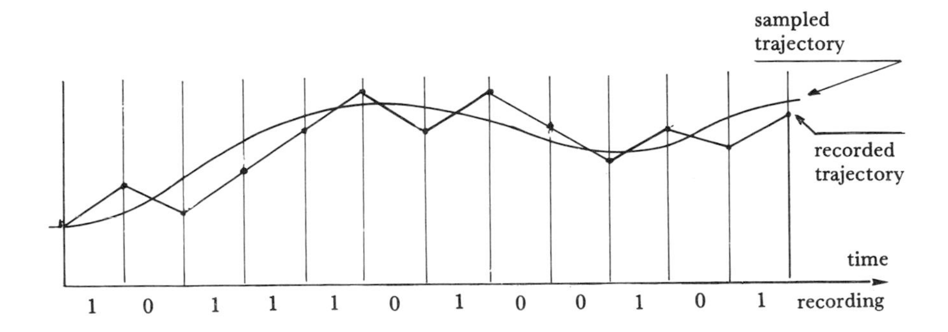

Figure 3.19. *First-order binary recording*

In each of these two basic methods, a certain amount of filtering during reconstruction is necessary, but this may sometimes be carried out mechanically by the robot. The errors produced in such methods are minimal and the rate of condensing may be as high as 12 (where 1 bit can replace a word of 12 bits). Moreover, the errors are usually minimized by the human operator, particularly if he is telecontrolling the operation, that is, seeing the effects of condensing the data in real time.

In all cases of continuous trajectory recording, the series of points obtained form parallel lists (one list per DOF), which are divided into sublists or blocks. In each of these blocks, the information is stored sequentially. Depending on the system, there may be possibilities for creation, insertion and deletion of data, as in point-to-point control robots.

When a complete program for a continuous path robot cannot be stored in the central memory because it is too long, it is recorded on a diskette, tape or hard disk, and at the moment of execution it is transferred in blocks to the central memory using a buffer: during execution of one block, the next block is loaded into the central memory in an empty space or a space occupied by an already executed block.

3.4.5 PROGRAMMED TRAJECTORIES

Programmed trajectories are not recorded during training but they are programmed symbolically using procedural instructions (as in a computer). The memory must store these instructions, and also data which may relate to positions, as in the previous example, and to numerical values which will affect the execution of the program. This type of programming is described in detail in Chapter 4.

3.5 Master control unit

In all servocontrolled robots, a master control unit is required to transform the static data contained in the memory into dynamic commands to be given over a period of time to the positional servo-systems in the various DOF. In the first flexible robots, which were point-to-point trajectory type the master control unit is based on the simple principle that the list of points recorded is entirely sequential (both logically and physically in the input medium) and the conditions for moving from one point to another were extremely simple (timing and sometimes external signals). In this case, a unit wired specially for this type of robot and its associated memory could be formed using logic circuits.

The development and decreased price of microprocessors have made this technique obsolete. All servocontrolled robots (and even some pick-and-place robots) currently being developed are sequenced by one or more microprocessors. Each microprocessor is associated with software developed by the robot designer, so that the unit can carry out the task or tasks required. The software often directs the following functions, simply by changing the program:

1. *training*: the system manages the man-machine dialogue and stores the information for the task program in its memory;
2. *program management*: with this function the user can edit programs (eg modify, place them in order, store them on external input media, load them into memory, delete them).

   It should be noted that some robots are equipped with a facility that operates away from the factory floor, for example, by telephone link (eg modem) between two computers. In such cases, the manufacturer can study what is going on in a machine far away, by using another system;

3. *execution*: the processing system guides the movement of the robot.

This function will now be considered in detail.

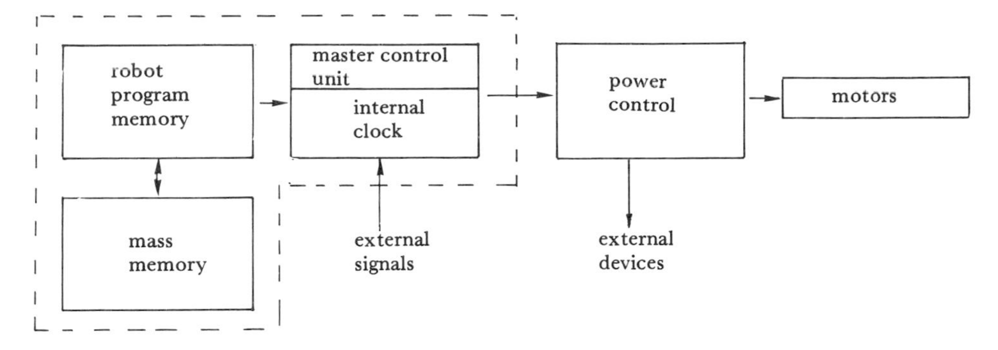

Figure 3.20. *Processing system*

The robot processing system is shown in Figure 3.20. The function of the master control unit is the generation, over a period of time and according to the task program and external signals, of the commands to the drives that power the robot. Functionally, the task of sequencing is presented in Figure 3.21, where only the sequencing and servocontrol functions are always present, irrespective of the type of robot (programmable servocontrolled). The two other functions of trajectory calculation and coordinate transformation depend on the degree of sophistication of the robot and its applications, but they are frequently included in modern robots, and used for increasingly varied applications. On the operational level, these different functions are carried out in a variety of ways, because of the potential offered by multiprocessor systems. It is possible to choose the architecture best suited for each function according to the performance standards required (see Figure 3.22).

In Section 3.5.1 these logic functions and standards of performance are described.

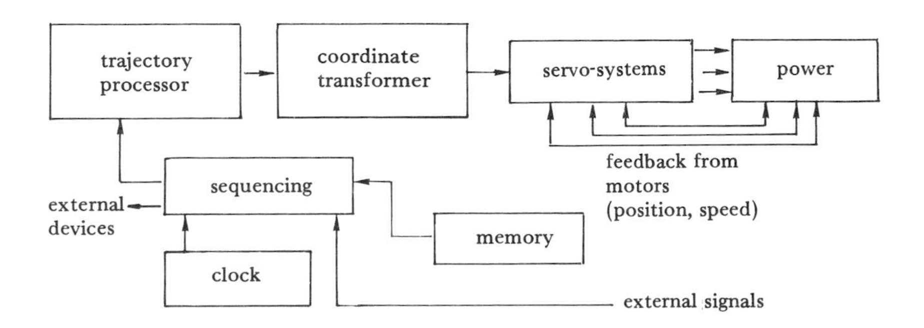

Figure 3.21. *General master control system*

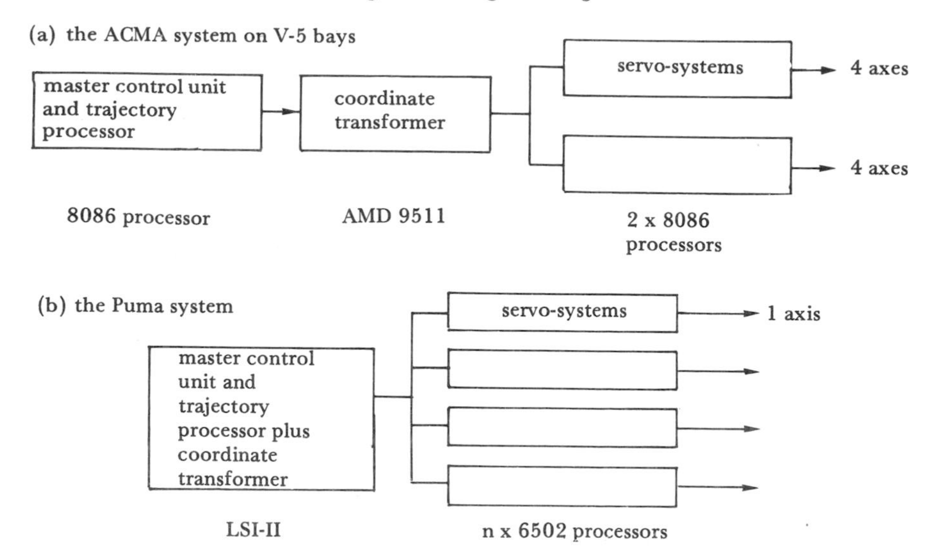

Figure 3.22. *Processing architecture: (a) the ACMA system and (b) the Puma system*

3.5.1 SEQUENCING

This function is, without doubt, the simplest to carry out, since the computer is only required to find simple instructions stored in the memory, to interpret them and activate the corresponding actions at the appropriate times. The simple instructions (eg go to a given point, generate a given trajectory, wait for a particular combination of signals) are stored in the memory either sequentially or in chains (ie each instruction indicates the location of the following instruction in the memory).

The addressing facility of processors allows simple access to the following instruction. If the entire program cannot be stored in the central memory, the master control unit must read into central memory the next block so that it is available when its first instruction must be executed. The timing of instructions is triggered by the end of the action in progress. The end can be shown by:

1. an internal delay fixed by the designer for each type of instruction;
2. the arrival of the robot at its destination point with a degree of precision which can or cannot be fixed by the user;
3. a combination of external signals.

Sequencing programming language type instructions is described in Chapter 4. It should be mentioned that although the series of instructions can be stored in a line, the sequence of execution is not ncccssarily linear: branching instructions (either conditional or unconditional) can be used to describe algorithms (eg in palletization where not all the positions are stored in memory).

3.5.2 CALCULATING TRAJECTORIES

Calculation of a trajectory is necessary each time the robot must move from one recorded point to another under some constraints. Without this, the robot trajectory between two recorded points is unpredictable (cf point-to-point trajectories, Section 3.4.2) and sometimes differs from one execution to another. In this situation, the trajectory depends, in space and time, on the characteristics of each servo-system, which may vary with time (differences in temperature, friction coefficients, load etc). When the points are very close together, as in continuous trajectories, the problem does not arise but when the points are widely separated (as in point-to-point control robots). The inaccuracies in the trajectories may impose the introduction of intermediate points. When the constraints on the trajectories are relatively simple, it is usually possible for a calculation module to generate the closer intermediate points which will guarantee that the trajectory is followed with precision, as is the case with continuous trajectory robots.

The simplest example is that of the coordination of the DOF. Consider a robot that must go from a configuration P_1 (given by positions $\alpha_1^1, \alpha_2^1, \alpha_3^1, \ldots, \alpha_n^1$ of its n DOF) to a configuration P_2 (given by $\alpha_1^2, \alpha_2^2, \alpha_3^2, \ldots, \alpha_n^2$) in time t. The coordination of the DOF will generate the intermediate positions, allowing all the DOF to reach their final position simultaneously. The time t taken for movement from P_1 to P_2 may either be specified by the programmer, or calculated so as to be as small as possible. In the latter, and more frequent, case, the calculation module must generate the minimum time required for each axis to move from configuration α_i^2 to configuration α_i^1, while taking into account the mechanical constraints of each axis (generally maximal acceleration $\ddot{\alpha}_{i\,Max}$ and maximal speed $\dot{\alpha}_{i\,Max}$) (see Figure 3.23).

Depending on the size of the trajectory, two speed profiles can be obtained (see Figure 3.24).

Let $\delta_i = \dot{\alpha}_{i\,Max}/\ddot{\alpha}_{i\,Max}$ be the time required to reach speed $\dot{\alpha}_{i\,Max}$ at constant acceleration $\ddot{\alpha}_{i\,Max}$ and $T_i = (\alpha_i^2 - \alpha_i^1)/\dot{\alpha}_{i\,Max}$. If $T_i > \delta_i$, a constant speed plateau is necessary and the following equations of motion are established:

For $t \epsilon [0, \delta_i]$:

$$\alpha_i(t) = \alpha_i^1 + \tfrac{1}{2}\ddot{\alpha}_{i\,Max}t^2 \quad (3\text{-}5)$$

For $t \epsilon [\delta_i, T_i]$:

$$\alpha_i(t) = \alpha_i^1 + \dot{\alpha}_{i\,Max}\left(t - \frac{\delta_i}{2}\right) \quad (3\text{-}6)$$

For $t \epsilon [T_i, T_i + \delta_i]$:

$$\alpha_i(t) = \alpha_i^2 - \tfrac{1}{2}\ddot{\alpha}_{i\,Max}(T_i + \delta_i - t)^2 \quad (3\text{-}7)$$

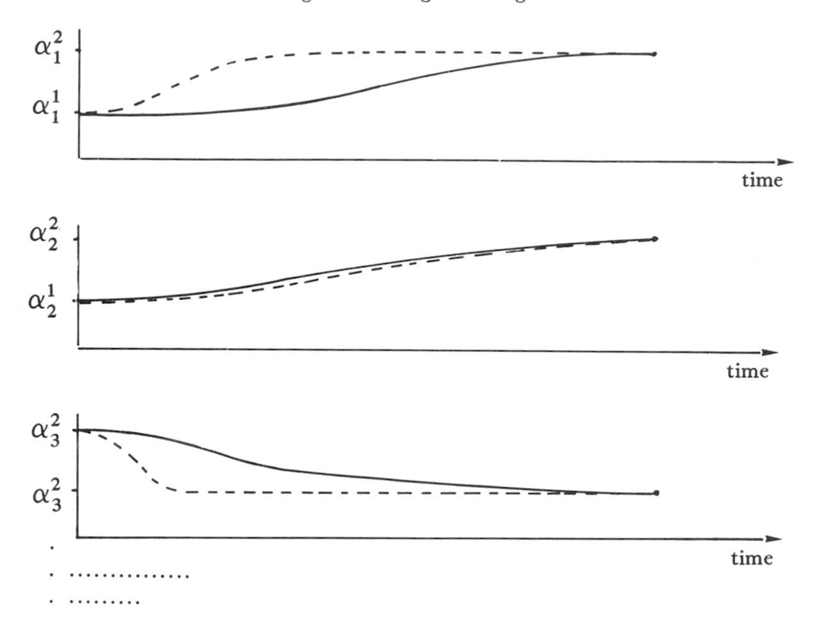

Figure 3.23. *Trajectories with (——) and without (- - -) coordination of the DOF*

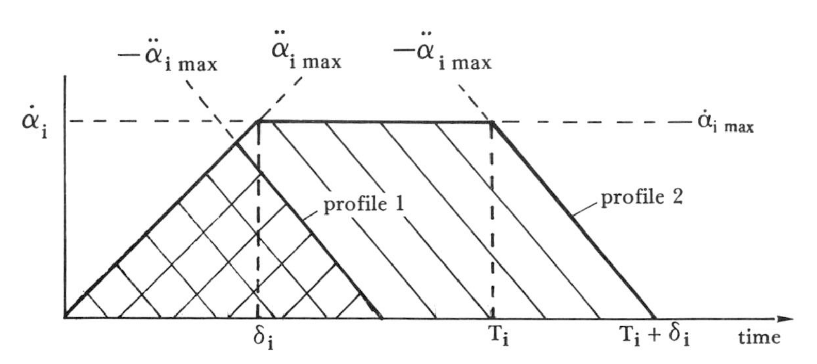

Figure 3.24. *Speed profiles for optimal speeds*

NB 1: depending on the type of actuator, deceleration may have an absolute value, different to that of the acceleration.

NB 2: the cross-hatched surfaces correspond to the desired movements of the actuators $|\alpha_i^2 - \alpha_i^1|$.

The duration of the minimal transition for axis i will then be:

$$\lambda_i = T_i + \delta_i \tag{3-8}$$

If $T_i < \delta_i$, the maximal speed is not reached and the duration of the trajectory is:

$$\lambda_i = 2\left[\frac{\alpha_i^2 - \alpha_i^1}{\ddot{\alpha}_{i\,Max}}\right]^{1/2} \tag{3-9}$$

For $t \in [0, \lambda_i/2]$:

$$\alpha_i(t) = \alpha_i^1 + \tfrac{1}{2}\ddot{\alpha}_{i\,Max}\,t^2 \tag{3-10}$$

For $t \in [\lambda_i/2, \delta_i]$:

$$\alpha_i(t) = \alpha_i^2 - \tfrac{1}{2}\ddot{\alpha}_{i\,Max}\,(t - \lambda_i)^2 \tag{3-11}$$

The displacement time will be equal to the maximum of the times λ_i, and the speed curves of the axes not constrained by this time will be modified so that all the axes terminate their movement simultaneously. There is an infinite number of ways in which an axis can arrive at its final position in a given time, if the time is not the minimum imposed by the mechanical constraints.

The trajectory calculation module must use a general method to generate automatically the laws of motion for the non-critical axes. The most commonly used method is the *proportional method* on the constraining axis. In this method, all increases in position are proportional to each other, which results in an extremely simple algorithm: the constraining axis acts as a model and all the other axes are guided proportionally. This also relates to the fact that the speed laws follow the same profiles. This method is not infallible: if, for example, as in Figure 3.25 two axes 1 and 2 have similar minimal durations of motion but significant differences in acceleration and maximal speed, it is clear that axis 1 will not reach its final position with the same speed profile as axis 2, with each one respecting its own acceleration constraints.

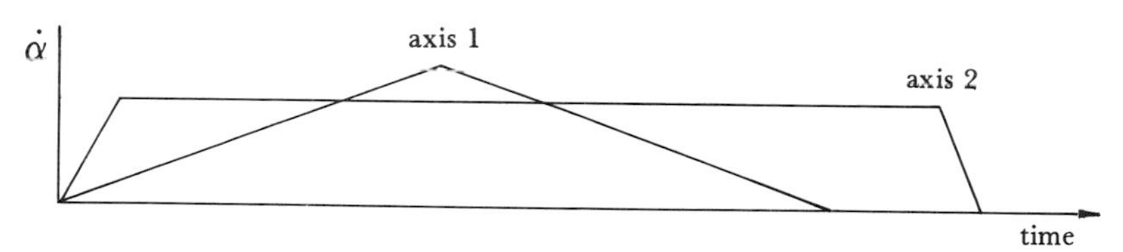

Figure 3.25. *Two speed laws where the proportional method cannot be applied*

The exact proportional method is more difficult to apply: it is necessary to calculate the durations of the three phases: acceleration, constant speed and deceleration, which satisfy all the constraints, while adding to a minimal value. Moreover, this method, although it retains the advantage of proportional increments (which makes application simple once the three phases have been calculated), it may lead to movement times greater than the minimum.

An excellent method for satisfying all constraints consists in choosing, for each axis, a law of motion similar to its optimal law, but extended in duration so as to be as long as that of the constraining axis. The application of this method is relatively arduous (both before and during execution) and does not retain the advantage of proportionality in movements (this makes the intermediate positions difficult to determine using an off-line system, for example, a CAD system).

More complex constraints than the simple coordination of DOF may be imposed on a point-to-point trajectory. This is the case, for example, when the trajectory of the tool must be a straight line (the expression used is 'linear interpolation'). The problem is relatively simple if the robot is Cartesian, that is, if the three first DOF are translational (generally with axes perpendicular, but this is not always the case). In this situation, the proportional method itself gives rise to linear trajectories for all points linked in a fixed way to one of these three axes. In the set of coordinate axes X, Y, Z of the robot, the proportionality constraint imposes the following equation:

$$\frac{\Delta X}{a} = \frac{\Delta Y}{b} = \frac{\Delta Z}{c} = f(t) \tag{3-12}$$

which is the equation for a straight line (and remains a straight line even if the set of coordinate axes is not perpendicular). If, on the other hand, the point under consideration is linked to the arm X, Y, Z by extra DOF (eg with rotations at the wrist), and if at least one of these DOF is actuated during motion, at that moment any point which is not on the axis of rotation no longer has a linear trajectory. This situation arises, for example, in continuous welding, using an arm with Cartesian axes. As long as the motions of the wrist are small in relation to the linear movement, the trajectory can be considered linear. If, on the other hand, the trajectory imposes significant changes of orientation for the torch over short distances, linearity of the trajectory cannot be guaranteed when the proportional method is used for the axes (see Figure 3.26).

In this situation, the empirical solution consists of introducing intermediate points into the trajectory to minimize these effects caused by the influence of the rotations.

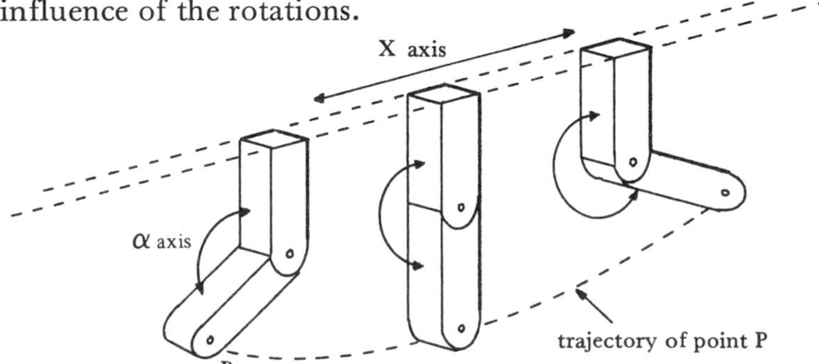

Figure 3.26. *Example of a proportional coordinated trajectory (one translation, one rotation)*

Another problem which is just as difficult to solve (and is often encountered in continuous welding) is that of the trajectory imposed at a given speed. The proportional method does not necessarily lead to a constant speed throughout the trajectory of the tool, even if a rectangular speed profile is imposed, thus guaranteeing constant speeds for each DOF. As in the previous example, the only case in which the speed is constant is that in which the only DOF set in action are either linear, or with an axis of rotation that passes through the tip of the tool. In the case of a Cartesian robot in which the movement imposes increments ΔX, ΔY and ΔZ on the three axes, if the speed imposed is V, the time T required to carry out the movement is given by:

$$T = \frac{\sqrt{\Delta X^2 + \Delta Y^2 + \Delta Z^2}}{V} \qquad (3\text{-}13)$$

which leads to motions being carried out at constant speed on each axis, as expressed below:

$$V_X = \frac{\Delta X}{T} \qquad (3\text{-}14)$$

$$V_Y = \frac{\Delta Y}{T} \qquad (3\text{-}15)$$

$$V_Z = \frac{\Delta Z}{T} \qquad (3\text{-}16)$$

In other cases (non-translational axis movements), the proportional method cannot give rise to linear trajectories, nor can it guarantee a constant speed.

For these reasons, but also to resolve other problems, such as the generation of more complex trajectories (eg circular), the translation of trajectories or the execution of simulated trajectories (eg generated by CAD), the solution consists of expressing the trajectory in a convenient set of coordinate axes, independent of the robot, which will generally be a set of Cartesian axes linked to the task. The calculation of the trajectory in this set of coordinate axes is easily made and the calculation module of the robot will simply have the task of generating the intermediate points in the Cartesian axes which will satisfy the constraints imposed. These points will be translated as they are generated (in a time which will be shorter than the period of generation of the Cartesian points) into robot configurations, using a coordinate transformer.

3.5.3 THE COORDINATE TRANSFORMER

The function of the coordinate transformer is:

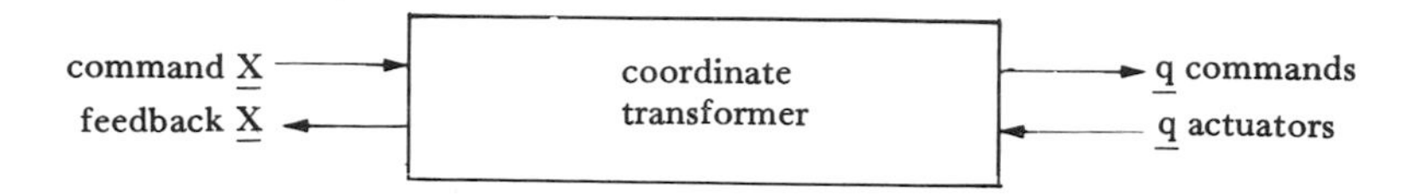

The tool is defined in a set of fixed Cartesian axes 0XYZ by a vector $\underline{X}$ with six components formed by the coordinates of point P of the tool in question (TCP or Tool Centre Point) and the orientation of a trihedron R linked to the tool (determined, eg, by its three Euler angles):

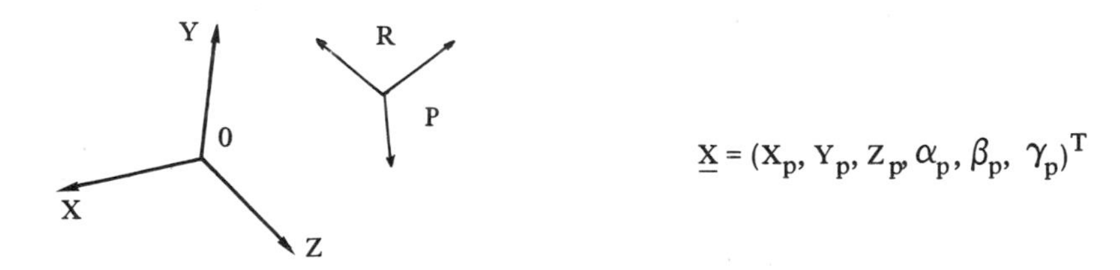

$$\underline{X} = (X_p, Y_p, Z_p, \alpha_p, \beta_p, \gamma_p)^T$$

1. The robot is defined by a vector $\underline{q}$ with n components derived from the values of n articulated variables (rotations or translations).
2. The two vectors $\underline{X}$ and $\underline{q}$ are linked by a non-linear relationship, usually complex but easily obtained in any situation (programs have even been written to generate this relationship automatically for an arbitrary robot defined by its geometric characteristics):

$$\underline{X} = F(\underline{q}) \quad (3\text{-}17)$$

This relationship is called the *direct geometric model.* The problem is to find the inverse relationship:

$$\underline{q} = F^{-1}(\underline{X}) \quad (3\text{-}18)$$

that is, to determine the commands to be sent to the various DOF so that the tool reaches the desired configuration $\underline{X}$ expressed in the task set of coordinate axes (usually a Cartesian coordinate set).

This problem which is described in detail in Volume 1 of this series entitled *Modelling and Control* (Coiffet, 1983) is usually complex since a number of other problems are encountered. These are described below.

1. The existence of a solution (S). Unlike the direct geometric model, the inverse model may not always have a solution (if the position cannot be reached by the robot), may have several solutions (the robot may take up a number of configurations for a given value of $\underline{X}$) or may have an infinite number of solutions (a robot with more than six DOF or with degenerated configuration, as in a situation where rotational axes are aligned).

2. *The analytical expression of the solution.* The analytical solution of the inverse of F does not necessarily exist, particularly for robots with complex geometric configurations (eg wrists with three non-intersecting axes). The inverse kinematic model which is, in general, easier to obtain can be used to bypass this problem. It is based on the Jacobian J of the equation F:

$$\Delta\underline{X} = J\Delta\underline{q} \longrightarrow \Delta\underline{q} = J^{-1}\Delta\underline{X} \tag{3-19}$$

3. *The calculation time for transformation.* The calculations are usually complex and involve a large number of operations and trigonometric calculations. If specific precautions are not taken, the calculation timc for transformation may be prohibitive and quite incompatible with the on-line generation of trajectory points (in the first robots with coordinate transformers at Stanford, the calculations were made off-line as the program was compiled). With robots currently in use, the calculations can take between 10 and 100 ms, which is sometimes too long in relation to the performance of the robots, and can result in very irregular movements (one point every 100 ms at 1 m/s gives one point every 10 cm). In such a situation it would be necessary to generate intermediate points in the robot work space, using the proportional method. The methods used to reduce the time required for calculation are as follows:

- optimization of equations to minimize calculation time,
- calculations carried out with fixed decimal point (floating operations take far longer),
- tabulation of trigonometric functions,
- use of calculation processors (eg 8087),
- use of specialized processing structures (slice processors, multi-processors).

4. *The precision of calculation.* Because of the digital technology used (fixed length words, calculations with information loss) and the large number of calculations required, it is not unusual to obtain limited precision only during transformation from a position in the Cartesian space to a robot configuration, and vice versa. This is particularly noticeable during a 'back and forth' operation through a transformer: a robot configuration is transformed into a Cartesian position during training, then this position is reconverted to control the robot at execution time. The difference may be quite noticeable (on Puma systems, the user is advised to record the more vital points in angular form precision point rather than in their Cartesian form, so as

to avoid surprises). As the geometric model is only an approximation of the real robot, errors can accumulate, based on other errors. It is, therefore, often necessary to accept a certain degree of imprecision between the modelled positions and the real ones in the task set of coordinate axes.

5. Handling the mechanical constraints. It may happen that the solution of the inverse model imposes an order which is beyond a mechanical limit. If software limits are positioned within the mechanical limits, the coordinate transformer will notice the impossibility by comparing them and act accordingly. There are three possible solutions to this problem.

1. Movement can be interrupted and an alarm generated. Depending on the situation, this alarm may or may not be processed by the program. In the latter case, manual intervention is necessary.
2. Movement can be continued whilst constraining the endangered DOF until the order returns to the permitted zone. The constraints imposed on motion (eg linear motion or constant orientation of the end effector) will no longer be respected.
3. Possibilities can be examined so as to ascertain whether another configuration would allow the desired position to be reached whilst satisfying the constraints of the DOF. This solution may easily be applied to the rotational axes with a stroke of 360° or more: a position outside the limits must correspond to another position of the actuator within the limits of its rotation of ± 360°. Unfortunately, this method gives rise to considerable faults in continuity which, generally speaking, disturb the desired trajectory.

The trajectory calculated in the Cartesian space does not take into account the dynamic potential of the robot axes. The maximal speeds and accelerations are usually imposed on the Cartesian trajectory, but unless the motion is simulated, it is difficult to know whether or not the constraints will impose speeds and/or accelerations which will be impracticable at axis level. This means that the duration of trajectories in a space other than that of the robot are not optimized and/or that major deviations in the trajectories will occur when high speeds are required (see Figure 3.27).

The techniques for real-time coordinate transformation for robots have only recently been developed, and have only been used in robots at the top of the range since 1980 (even though Vicarm, the forerunner of the Puma systems, has used them since 1974). New processing architecture allows these techniques to be included at a reasonable cost and with acceptable performance standards. As shown previously, there are many advantages:

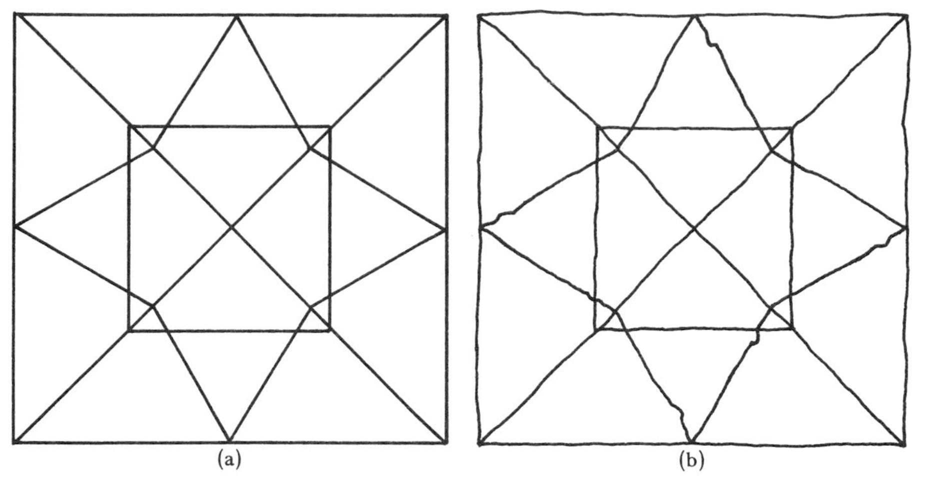

Figure 3.27. *Drawings made by a six DOF articulated robot equipped with a coordinate transformer with linear interpolations (from Citroen): (a) speed factor 0.1 m/s; (b) speed factor 0.8 m/s*

1. the robot can be controlled in the Cartesian mode and in relation to the tip of the tool (considerable reduction in the training time in the point-to-point mode);
2. trajectories can be generated with constraints (linear or circular interpolation, imposed speed etc);
3. an entire program can be translated to carry out a relocating operation or to follow a mobile target etc;
4. programs can be generated automatically using a CAD workstation;
5. successive positions can be calculated without all positions in a sequence being taught (eg palletization).

In the future, coordinate transformers will be sufficiently developed to allow them to be integrated into feedback systems, which will impose calculation time below a millisecond, and allow calculations, not only based on motion but also on force, to be carried out. The problem of force calculation is being encountered increasingly in, for example assembly and machining, but has not yet been satisfactorily solved.

3.6 Programming by training

At the present it can be stated that the most commonly used programming method for servocontrolled robots is that of direct training of the robot, installed in the site where it will finally be used. This method is called 'teaching by doing'. Since the introduction of the first Unimate systems, this method has been widely used in many applications. It is now being replaced in cases which are either complex, in which the training time is too long (in which case programming is carried out using a language), or when the production run will be too short to justify the training time (this problem is solved by the use of language, but also with automatic generation of training-type programmes using CAD/CAM systems). All methods of programming by training are based on a physical demonstration of the task to be carried out, by an operator on the site at which the robot will operate. There can be slight differences in man-machine dialogue (from the ergonomic point of view), depending on the type of robot.

3.6.1 POINT-TO-POINT AND CALCULATED TRAJECTORY ROBOTS

For these two types of robot, the human operator must move the robot into configurations which are recorded. There are two methods for this:

1. manual movement;
2. telecontrol.

3.6.1.1 Manual movement

In the first of the two methods the human operator manually moves the robot so as to bring the tool or gripper into the desired configuration. Recording is activated by the human operator who throws a command switch or uses the terminal keyboard, if the computer is equipped with one.

Clearly, such a method can only be used if the AMS allows motion to be brought about by direct action on the axes and not only by the motors. A robot that allows this type of motion is described as *reversible*. This is not possible if, for example, the demultiplication ratios are too large or if the transmissions are non-reversible (lead screws). In addition, the motor itself must be capable of being moved by its output axis. Although this is possible with electric motors, hydraulic motors require the use of a bypass, allowing communication between the two lines leaving the servovalve (see Figure 3.28).

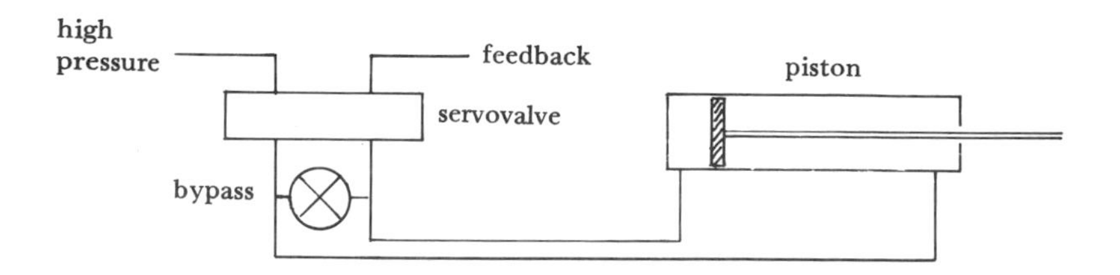

Figure 3.28. *Bypass in an hydraulically powered motor*

The final condition for manual movement is that the forces required to move the robot must be compatible with the physical strength of the human operator. This implies that the robot must either be light or well balanced (at least in its first three DOF).

Even if all these conditions are satisfied, manual movement, carried out without any power being transmitted to the end effectors, may not be very precise. The positions of the various end effectors recorded without any power do not generally correspond to the positions necessary for attaining the desired configuration when the motors are controlled. A number of factors contribute to the fact that that same equation does not apply:

$$\underline{X} = F(\underline{q}) \tag{3-20}$$

when the robot is controlled by the actuators and when it is inert and placed in position by the human operator. Among these factors are:

1. degree of play in the transmissions;
2. flexibility of the transmissions and the rigid sections;
3. errors of the servo-system.

The manual movement method is, therefore, not generally used except with small robots and when the degree of precision required is not very great.

3.6.1.2 Telecontrol

In other situations the robot is moved by its actuators which are controlled by the human operator. Control is generally established using a control box, often called the *teach pendant* connected by a cable to the robot master control unit. In most cases, the control box is equipped with an array of switches which allows the robot's DOF to be actuated one by one. A number of other commands are included, allowing the speed to be chosen, for example, or dialogue to be carried out with the master control unit, for adjustments to be made to the program (program editing). To improve this dialogue, the teach pendant is sometimes equipped with a screen of one or more lines, on which messages can be displayed (see Figure 3.29).

In the majority of cases, each DOF is controlled by switches. These may take the form of two keys for each of the directions of actuator movement, or a switch with three positions. These switches or keys activate movement of the actuator at a constant speed which can usually be modified by another command (potentiometer, encoding disk or scale factor). The coordination of several DOF is virtually impossible, no matter which method is used, so the human operator works sequentially on one DOF at a time.

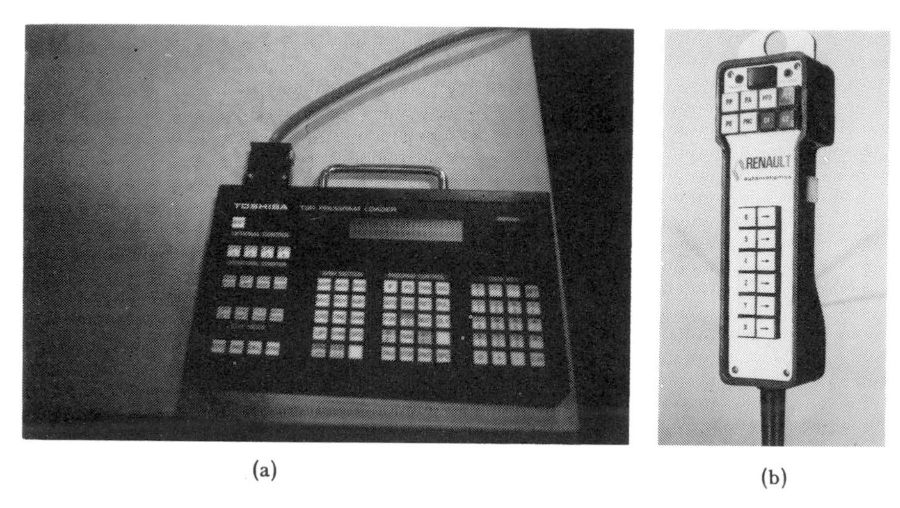

(a) (b)

Figure 3.29. *Teach pendants: (a) Toshiba; (b) ACMA*

The technique of moving axes using binary switches is extremely difficult to master. When one or more of the DOF are rotations, in particular, the task of the human operator is far more complicated and it is quite common for even an experienced operator to take several minutes to reach the desired configuration. This problem is well known to telecontrol specialists, who have known for a long time that of all the methods for controlling a robot, binary control is the least satisfactory. (The best method is that of positional control with force feedback, or master-slave control.)

To improve these performance standards, designers of industrial robots are developing the use of coordinate transformers, described previously. In this method, the switches directly control the movements of the tip of the tool in Cartesian space. The main advantage is that the movement for the orientation of the tool is separated from the overall trajectory movement of the tool tip. Performance standards are significantly improved, and reduction in programming time of factors of between 5 and 10 can be achieved.

A slight improvement on this method has been made to a number of recent robots, by the use of a *syntaxer* or *joystick* which allows, with a little practice, real control of the robot to be carried out, by acting simultaneously on several DOF. Control is generally better than with simple binary switches, with linear or exponential functions with a neutral zone (to avoid a slow drift of the robot in the lever's neutral position) for each DOF X.

A syntaxer generally has three DOF (two syntaxers are required to control a six axis robot) but there are prototypes using six axis control.

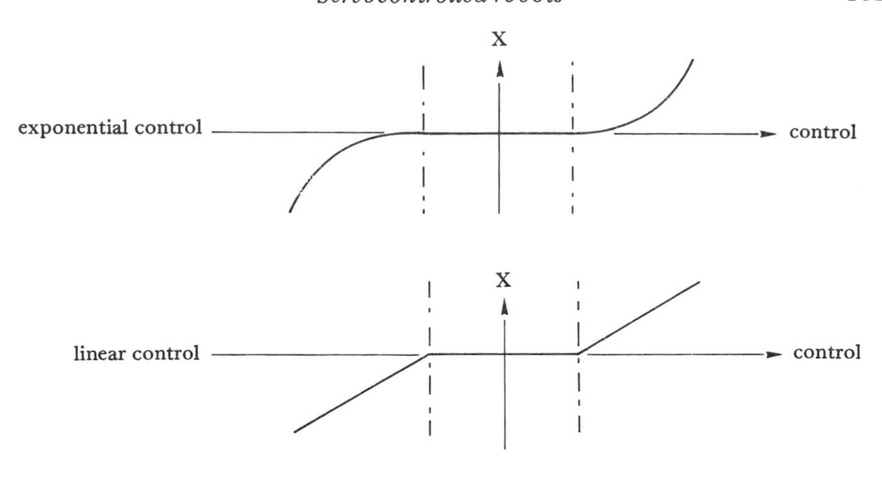

Figure 3.30. *Control laws for axis movement*

Figure 3.31. *ASEA teach pendant with syntaxer*

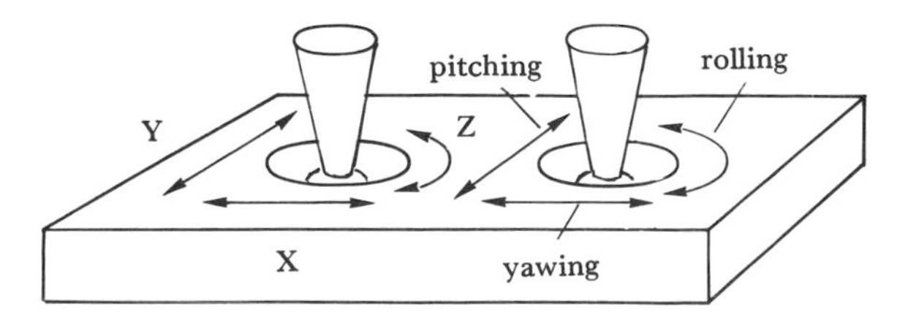

Figure 3.32. *Syntaxer with two joysticks which control three DOF each*

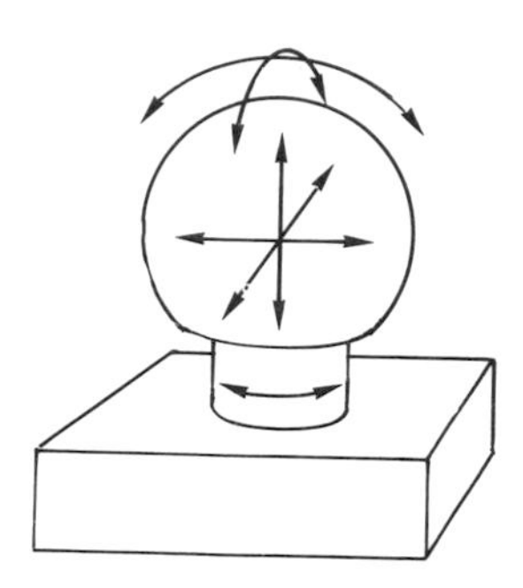

Figure 3.33. *Syntaxer with six DOF*

Control by syntaxer is generally performed in an absolute set of Cartesian axes and in relation to the tip of the tool. When different tools are used, the tool transformation must be changed to accommodate the new relationship between the wrist set of coordinate axes and the tool set of coordinate axes (see Chapter 4). This is carried out using a keyboard, which may only be numerical.

It is possible, however, to control the tool in its own coordinate set: the tool advances, moves back, rises, descends and moves to left and right in relation to itself. This is often used in telecontrol, as, for example, in the SPAR arm in the space shuttle.

The human operator controls the tool as if he was in its place (in the same way as an aircraft is piloted by telecontrol). In some cases, human operators find this method of control more convenient than control in an absolute coordinate set. The ideal (and this does not present any problems for the master control unit) is to be able to choose the method used.

3.6.1.3 Ancillary control

In the two previous methods, the human operator is required to provide the master control unit with additional information to allow the program to be set up:

1. recording of a point;
2. nature of the point (terminal or intermediate);
3. external signals to be received or activated;
4. timing;
5. speed of movement;
6. the nature of the trajectory;
7. any sequence to be eliminated or replaced (program editing).

The increase in potential of servocontrolled robots has lead to the development of teach pendants equipped with all the available functions, controlled by a large number of switches, lights, selectors etc, but training operators to use such complicated controls is lengthy and complicated. The new approach based on a keyboard and VDU makes full use of the potential of microcomputers. The program used for controlling the robot during training also guides the human operator during his task by offering a limited number of choices at any time and asking him to feed the information directly via the keyboard or through menus. This approach is starting to be widely accepted in the industry following a long period of hostility and setbacks, due to the unreliability of the first microcomputers and terminals used in industry. The widespread use of this approach for machine tools with numerical control has also contributed to the acceptance of the keyboard by human operators. Program editing can also be carried out during the training phase, for example, to correct part of the trajectory, but also to repeat certain movements which have already been recorded. The most sophisticated robots also use a 'palletization' function, which allows a whole sequence to be automatically generated from the teaching of a single element of the sequence. This type of programming leads the way to full feature programming languages as seen in Chapter 4.

3.6.2 RECORDED TRAJECTORY ROBOTS

When the robot must perform a more complex trajectory, which cannot be reduced to a small number of points or a curve described by algorithms of reasonable complexity, the only method of programming that can efficiently be used is to 'show' the trajectory to the robot. There are several methods by which this type of training can be carried out:

- manual displacement (assisted or not);
- the dummy arm;
- telecontrol.

3.6.2.1 Manual displacement

As with point-to-point control robots, the human operator moves the robot manually. In this example, however, he usually holds the tool itself and moves it at the desired speed along the desired trajectory. The mechanical parts of the robot follow the trajectory which is recorded by sampling the values associated with each DOF. When the mass of the robot is too great or when the actuators prevent the human operator carrying out his work, muscular assistance devices can be used, which take the form of a handle mounted on the tip of the robot and allow it to be controlled without difficulty (rather like power-assisted steering in a car). Unfortunately, the cost factor generally leads to the servo-system's being limited to the first three DOF and the human operator must lead the tip of the robot with one hand while directing the tool with the other, which is not easy to coordinate.

With or without assisted control, this method often proves awkward for the human operator, whose movements are hampered by the robot, and this is particularly the case for paint-spraying or projection systems. Moreover, as in point-to-point trajectories, repeatability between the recorded and executed trajectories is not good.

3.6.2.2 The 'dummy' arm

In this method, which can also be used with point-to-point control robots, the robot is replaced on site by a mechanical structure with identical geometry to that of the robot, but unmotorized and very light. The human operator is no longer hampered by the mass and actuators of the robot and his work is, therefore, greatly facilitated. He is still, however, faced with problems relating to the physical dimensions of the substitute mechanism since these are generally identical with those of the robot, so as to simulate any possible collisions.

The problems of repeatability are the same as before, with the added differences which may exist between the dummy and the robot (structure, sensors) and between the positioning of the dummy and that of the robot (which will be replaced on site for execution of the task).

3.6.2.3 Telecontrol

This method was developed to help solve the problems mentioned above. The robot, on its permanent site, is telecontrolled using a *syntaxer.* The syntaxer must allow the human operator to carry out the necessary movements on the robot without constraint. The syntaxer is usually a dummy with a structure almost identical to that of the robot (total precision is not essential, since the human operator can observe the effect of his commands on the robot directly. Control is established in the position mode as in telemanipulation, but without force feedback (this is not necessary if there is no contact).

The human operator can use the optimum position for his dummy so as to benefit from the best work conditions. The repeatability of the recorded trajectory and the executed trajectory is optimal because it is equal to the repeatability of the robot, with the human operator integrating even the errors caused by data compression.

The only fault which is sometimes found with this method is that the human operator is no longer in direct contact with the tool. This method may be extended for use with giant or microscopic robots in the future, since there can be an arbitrary scale factor between the dummy syntaxer and the robot (see Figure 3.34).

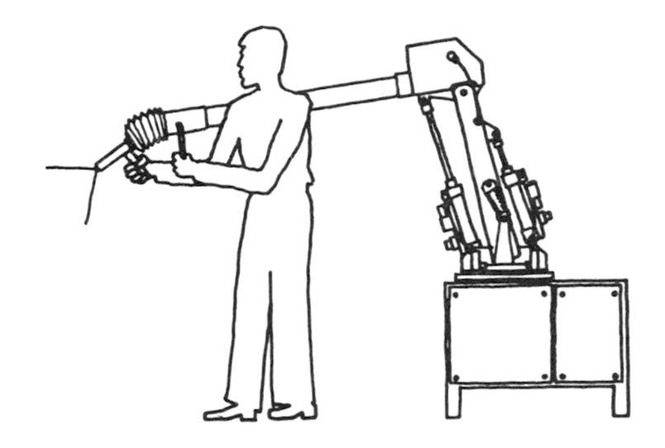

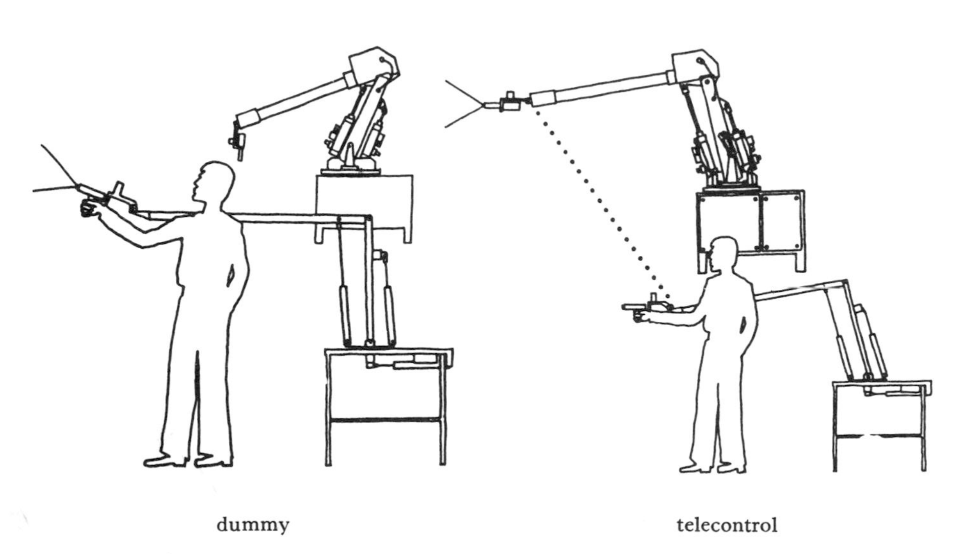

Figure 3.34. *Methods of continuous programming (from AKR)*

3.6.2.4 Ancillary control

As with point-to-point control robots, ancillary control methods are necessary to develop a program. The unit processed by the program editor is not, in this case, the point in space but an element of the trajectory. It is, therefore, necessary to define these elements, and this is effected by recording, using a recording button on the robot or dummy which defines an element each time it is pressed.

An element can generally be broken down into subelements. This

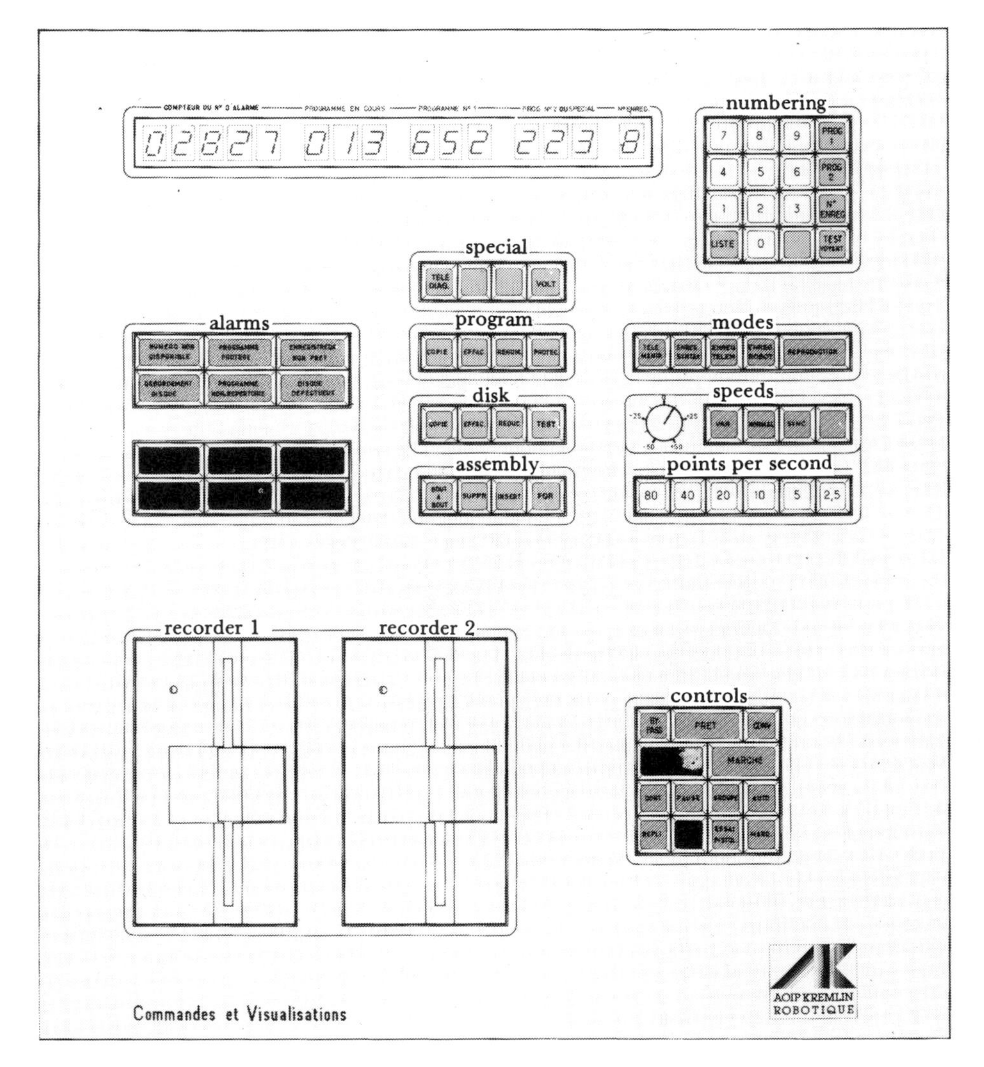

Figure 3.35. *Control panel (from AKR)*

allows program editing by eliminating some parts of the trajectory and replacing them with others. The master control unit automatically generates the 'connections' between the elements of the trajectory by calculating a movement completed in minimum time. These editing functions are very important and can make the difference between a good and bad recorded trajectory robot. It must be stressed that an optimum program cannot be obtained at the first attempt and sometimes a large number of adjustments must be made to allow the correct operation to be made in minimum time.

Apart from the editing function, the ancillary control methods are generally more limited than those of a robot with calculated trajectory. Standard control panels have, therefore, been used for a long time but screens are beginning to appear, and it seems that keyboards will follow (see Figure 3.35).

Chapter 4

Programming languages

4.1 Development

In parallel with the development of industrial robots programmed by training, as described in Chapter 3, other types of robot were produced in a number of laboratories, principally in the USA at Stanford University and MIT, and in Britain at the University of Edinburgh. These laboratory robots were the product of research, started in the mid 1960s, on artificial intelligence.

Artificial intelligence has never been precisely defined. The term is often wrongly used to mean that a machine takes into account modifications in its working environment or work space. In reality it concerns processing techniques which are intended to stimulate intuitive human reasoning, that is, the type of reasoning that cannot be formulated by equations or algorithms. Starting with this very different approach, some researchers hoped that they would create machines which would replace man in tasks normally considered impossible for computers, such as automatic translation, playing chess, scene analysis and exploration in unknown environments.

For the last of these tasks, it was necessary to design machines capable of reacting to their work space; thus a different type of robot was produced. Some, such as Shakey, were mobile robots capable of moving from one room to another and carrying out simple tasks involving furniture moving etc, whereas the majority were simply handling devices (see Figure 4.1).

The constraints imposed on these machines by the computer scientists were straightforward:

1. the machines must be able to move small light objects in a restricted work space and therefore be provided with six DOF;
2. the machines must be controlled by computer and therefore fitted with digital control;
3. the machines must be easily maintainable and therefore run off electricity;
4. the cost must be low, hence the simplicity of solutions and the sometimes poor performance.

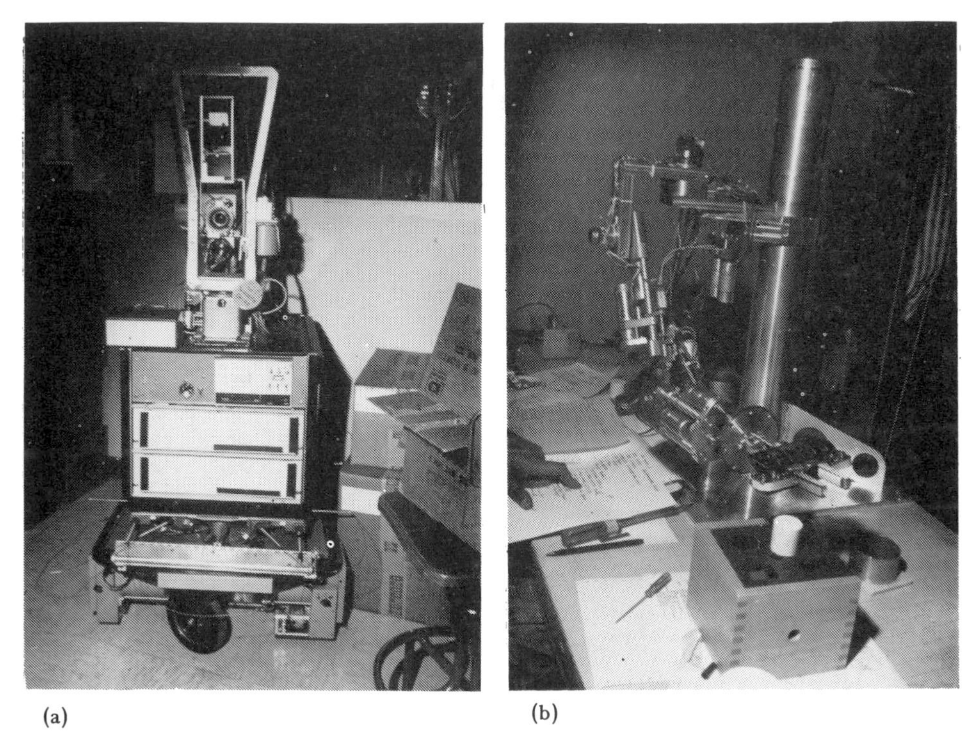

(a) (b)

Figure 4.1. *Laboratory robots: (a) Shakey; (b) Rancho*

Three of these manipulators (see Figures 4.1 and 4.2), developed during the late 1960s and early 1970s, enjoyed some success, and several copies were sold to both public and private laboratories:

1. the Rancho arm, developed from prosthetic devices;
2. the Stanford arm, developed and sold by Vicarm;
3. the MIT arm, developed and sold by Vicarm. (Vicarm was started by Victor Scheinman, a student at both Stanford University and MIT).

Although research into artificial intelligence and in particular into scene analysis and hand-eye coordination projects seemed to stagnate in the mid 1970s (and as a result the funding was reduced), research into computer-controlled manipulators based on traditional mechanics and programming made considerable progress from 1975 onwards. The processing methods used allowed systems with higher performance standards than those of the industrial robots of the time to be developed. A notable feature was the ease with which external events could be taken into account. The robots developed as a result of this research were referred to as *second generation robots*, to differentiate them

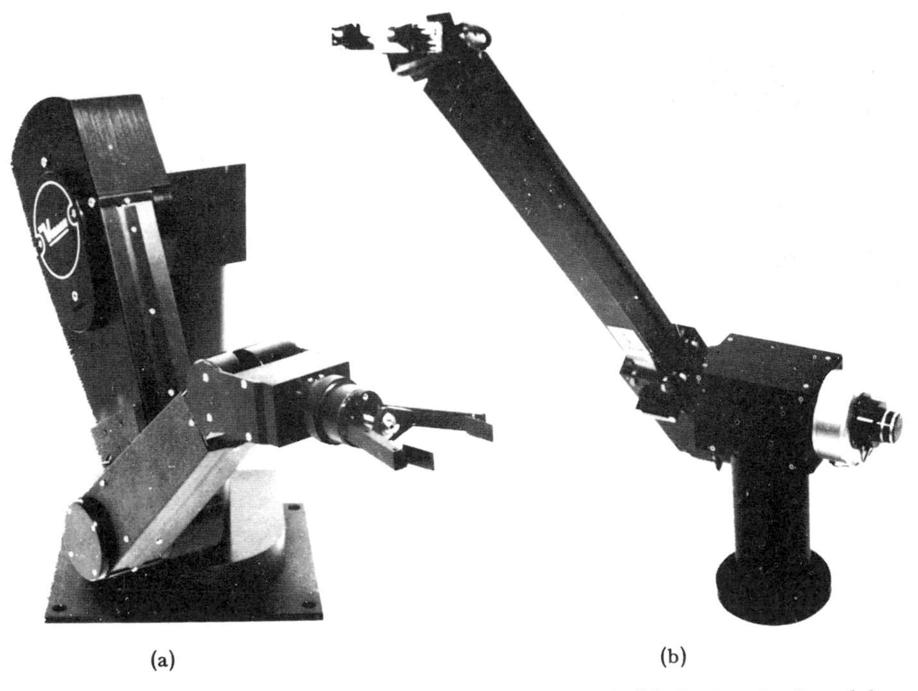

(a) (b)

Figure 4.2. *Vicarm manipulators: (a) the MIT model; (b) the Stanford model*

from *first generation robots* with purely repetitive cycles and no response to external events.

These second generation robots became the focus of interest with the introduction of the microprocessor, which provided additional processing power at a reasonable cost. In 1975–1976 the major manufacturers of robots began to experiment with these new programming techniques.

The first two industrial robots equipped with languages were launched on the market in about 1977. These were the Olivetti Sigma and the Unimation Puma.

In the Sigma, techniques of digital control were applied to assembly, but the SIGLA programming language was already using the capacity for reacting to information provided by the sensors (eg strain sensors) which would allow the detection of any malfunction in an operation and possibly allow the mistake to be corrected.

The range of Puma robots (Programmable Universal Machine Assembly) can be traced directly to the laboratories at Stanford University, since they are an industrial version of the Vicarm robots, after the company was acquired by Unimation. The VAL language and its processing architecture have hardly been modified at all. The Puma project was financed by General Motors, and was a sequel to another

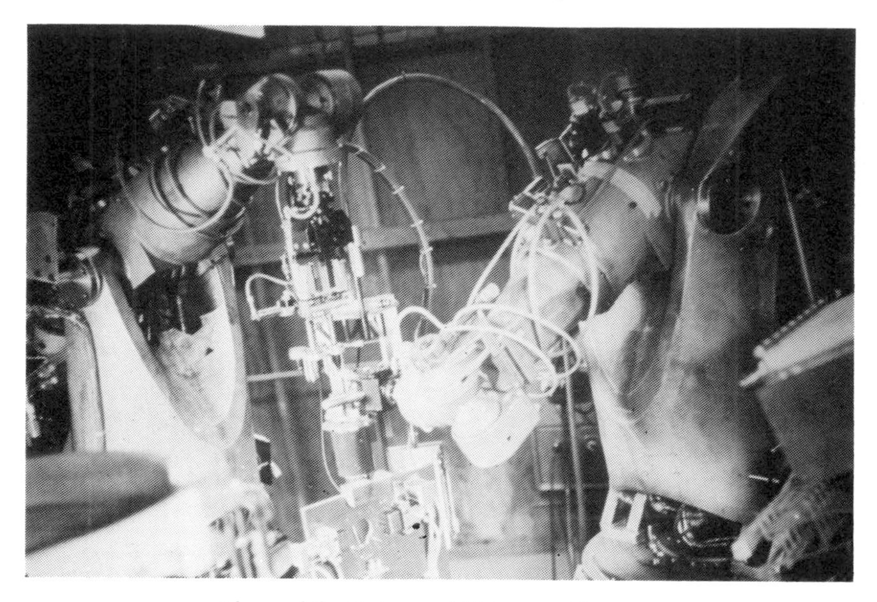

Figure 4.3. *Unimate 6000 assembly arms*

assembly project, undertaken by Unimation with Ford using first generation technology, which did not produce satisfactory results (Unimate 6000, see Figure 4.3).

These two precursors of the second generation robot experienced commercial problems when first put on the market. These setbacks were probably caused by the reticence of users towards the programming involved. It is also possible that the cost of these machines made them less attractive than manual labour. It was not until 1982 that second generation robots, and in particular language-programmable robots, were adopted by industry, with several hundred units in use all over the world. At this time also, many manufacturers, both well established and newcomers, started to produce language-programmable robots. Among the best known are Automatix with RAIL, Renault-ACMA with LPR, Scemi with LM, Sankyo with SERF, Microbo with IRL and IBM with AML. Then, in 1983 manufacturers of processing systems attempted to market robot control modules supplied with languages, but with no mechanical architecture. This approach was similar to that of the manufacturers of numerically controlled systems for machine tools (General Electric, Siemens, Fanuc, NUM etc) and it was intended for the many machine manufacturers lacking the expertise (or a sufficiently large market) to develop their own computer-based control. The first manufacturers to offer control stacks for robots with language were General Electric with PCL (a language intended for use with machine tools and robots), GIXI (affiliated to French CEA) with ROL and MATRA with the language LM on its Syscomat. However, a large number of manufacturers of controls are also offering products to robot manufacturers for standard operation of their hardware (point-to-point or continuous path in the teach mode).

Despite the fact that the world distribution figures are still very small, the potential for language-programmable robots is enormous because the areas to which they can be applied fall into a considerable number of different fields. The estimated growth rate for the years between 1983 and 1986 is more than 100 per cent and should remain above 50 per cent until the year 1990.

4.2 Programming levels

In Chapter 4 it was shown that modern servocontrolled industrial robots are controlled with the help of one or more microprocessors. For standard techniques such as those described in Chapters 1 to 3 (eg point-to-point, calculated trajectory, recorded trajectory) these microprocessors are programmed by the designer leaving the user with only a limited number of programming choices. If, on the other hand, the user could program the microprocessors, the robot's field of application could be considerably extended but at the cost of additional and significant programming problems. This is particularly true for the microprocessors used in robotics which are usually programmed in *machine code* (*assembler*) for reasons of performance, and to allow efficient management of real-time tasks (eg in servo-systems controlling several axes). This level of programming (known as level 0) is, therefore, never used in industrial robots. It is, however, used with special robots made on demand and for teaching robots connected to personal computers.

To avoid this complex and tedious form of programming, the first workers to use computers for controlling robots developed specialized software to allow far simpler programming. Unlike the *compiler approach*, usually chosen to simplify the programming of computers, in robotics the *interpreter approach* is used, in particular for solving problems of timing and real-time management. In this method specialized software, written once and with no further modification, decodes the instructions in the user program and controls the various motions of the robot. The user writes a program either by using the set of interpreter instructions directly, or by using a high-level language which is then translated into basic instructions. Depending on the power of the set of instructions placed at the disposal of the programmer, the language is classified in levels. Four levels of programming are recognized, which depend on the concepts handled by the programmer.

4.2.1 LEVEL 1: ACTUATOR LEVEL

The user can program displacements of the actuators either in a coordinated or uncoordinated manner. The time of displacement may, in some cases, be specified, otherwise it is automatically calculated by the interpreter so as to be kept to a minimum. This type of programming has been inspired by the numerical control languages used for machine tools. It is generally found in Cartesian robots (IBM 7565, Olivetti-Sigma) but it is also used in the simpler assembly robots (Sankyo, Sormel-Cadratic). The linear and circular interpolations are not generally supplied if more than three axes are concerned.

4.2.2 LEVEL 2: END EFFECTOR LEVEL

At this level the user programs movements not of the actuators but of the tool itself. This is particularly interesting for non-Cartesian robots and Cartesian robots when rotations of the tool are involved. It is possible, for example, to control the tip of a tool in such a way as to make it follow a straight or circular path whilst changing its orientation in the course of motion. This level of programming also allows the centre of rotation to be chosen for tool orientation motions and thus separates the displacement and orientation actions.

4.2.3 LEVEL 3: OBJECT LEVEL

The user is concerned with the motions of the robot but programs only the movements of the objects. This is known as *implicit programming* as opposed to *explicit programming* (eg levels 1 and 2). The motions of the robot are generated automatically on-line or during a compilation phase, by modelling the parts to be handled and the work space. This level of programming has been researched for a number of years, but as of yet is not used in any industrial robots. It must be added that this mode of programming can be as complex as explicit programming, it is not as generally applicable and it requires extensive modelling of the parts and the assembly area or work space.

4.2.4 LEVEL 4: OBJECTIVE LEVEL

This is a generalization of level 3. Only the final objective is specified and details of the intermediate stages are omitted. This level of programming is the subject of research and makes full use of the techniques of artificial intelligence with particular reference to the generation of *action plans* (see Volume 6 of this series entitled *Decision and Intelligence*). The difficulty is as much in the generation of these plans as in the specification language which must allow a description of the desired objective to be made without ambiguity (this is far from simple, even using natural language). The objective level, nonetheless, can be envisaged as becoming operational in the near future for relatively simple tasks but in work areas where unpredictable situations might arise or a high degree of autonomy is required, for example for programming mobile robots for exploration or intervention.

There are two basic approaches to robot programming languages in current industrial robotics (ie levels 1 and 2: actuator and end effector) which appear to be in opposition to each other:

1. the simplified but limited language approach;
2. the complex but powerful language approach.

Using the former method the designer is thinking of the final users, who are usually unfamiliar with computer technology. The number of instructions is, therefore, limited and the programming structure will be linear with very simple data structures. In general, this type of programming is no more effective than programming by training and the resources of the computer are underexploited.

Using the latter method all (or nearly all) the resources of the computer on the programming level are accessible to the programmer (but with varying degrees of ease depending on the language used). This programming power is, however, balanced by the complexity of the language, which requires extensive knowledge on the part of the user. This type of language is intended for specialists (eg system houses), who will generally not be the machine's final users. If each application is effectively programmed, the final users can have access to certain functioning parameters without having to know the robot programming language.

The division between these two approaches is often not well defined, with designers attempting to offer the maximum programming potential while not exceeding a level which is accessible to users untrained in computing. For the present, and probably for some years to come, these two objectives are irreconcilable. By careful examination of the programming capacity offered by the designer, the user can determine the category to which his robot belongs, and whether it corresponds to his requirements.

The programming languages available to users can be considered under four headings:

1. general programming facilities (independent of robotics);
2. facilities for modelling in geometrical space;
3. instructions for robot movement;
4. use of sensors specific to robotics.

Before entering into the details of these programming facilities, the general architecture of processing systems allowing this type of programming and the execution of programs is discussed.

4.3 Programming and execution

Whatever the level of programming used to describe the task to be carried out by the robot (or robots), at the moment of execution the control system will use a very different representation of the program to that specified by the programmer. The user program is, therefore, translated with a degree of complexity appropriate to the level of programming into a series of codes representing instructions and data. The robot and its control system may thus be considered as a pseudo-processing device executing a program in machine code. At present, this programming level in machine code is not evident to the user since the translation of the programs is generally carried out in the control system of the robot itself. Programming by language does, however, lead quite naturally to off-line programming (ie programming carried out without the use of the robot or its control system).

Under these circumstances, programming and translation can be carried out on a distinct and perhaps far more complex processing unit than the robot control system. If this approach is used, a single programming station can be used for several robots. The problem which has arisen with the appearance of the first robots using these techniques (eg Sankyo, IBM, PSA) is the definition of an intermediate code, and the decision whether to create a code of this type common to all robots. The supporters of this approach cite the example of the development of numerical control, with its many languages (eg APT, EXAPT, PROMO, PAM-ELAN), and the definition of an international intermediate code (CLDATA) easily translated by post-processors into executable code (often in the form of a punched paper tape) for any machine. A suggestion from Germany to establish an intermediate code (IRDATA), similar to the one mentioned above, but taking into account the specifications of manipulation, is being discussed in Europe. Other propositions from the USA and Japan are also being developed.

The state of robot programming techniques (and particularly those for second generation robots) is undergoing rapid development and it is impossible to define all the *basic instructions* which could one day be available for the more sophisticated robots. The suggestions concerning intermediate codes like IRDATA must, therefore, remain as open as possible (Table 4.1).

type of record	name	explanation
1,000	instruction source	The record contains the number and type of the user program instruction.
2,000	control data etc	The record contains technical information, such as acceleration, speed, tool number, input/output, duration of motion, conveyor tracking, palletization etc.
3,000	data	The record contains geometrical data in standard form: point, line, vector, surface, circle, cylinder, cone, sphere, parallelipiped, coordinate system, transformations etc.
5,000	position and orientation	The record contains details for the motions (position + orientation). It is possible to define several points and to specify a point-to-point or continuous path control (linear, circular interpolation etc). The displacements can be relative or absolute, precise or approximate.
9,000	unit of measurement	The record contains definitions of units of measurement (millimetres or inches – scale factors).
14,000	end of program	This record is the last in the program.
15,000	complex trajectories	The record describes the variations in a predefined trajectory. These variations can be saw-tooth, rectangular or sinusoidal.
17,000	tool description	The record contains the tool number, its geometrical parameters and its technological parameters.
19,000	robot description	The record contains parameters describing the robot.
20,000	sensor description	The record contains parameters describing the sensors.
21,000	Boolean and arithmetic functions	The record contains a list of operators in reverse Polish notation (the operations are carried out in a stack).
22,000	programming instructions	The record contains one of the many programming instructions (39 defined already), such as jump, call subprogram, pop and push the stack, waiting for signal, definition of new variables, definition of interrupts etc.
28,000 to 32,000	reserved records	These records are not standardized and are reserved for special applications.

Table 4.1. *IRDATA code*

– an IRDATA program consists of a sequence of records;
– each record includes between 2 and 125 words (integers, real numbers or alphanumerics) arranged as follows:

W_1	W_2	W_3	$W_4 \ldots W_n$
record number	record type	instruction code	parameters

4.4 General description

All processing languages are characterized by their capacity to define data structures, by the structures permitted by these data and by the algorithmic structure. For the level of programming closest to the machine (machine code or its mnemonic representation: *assembly language*), there is only one type of data: the *binary unit* (or bit) which has only two values (0 and 1). These bits are frequently grouped together, according to the wishes of the programmer, into *words* (generally of 8, 16 or 32 bits) which may represent *characters*, *integers*, *floating numbers*, *Boolean values*, *addresses of variables* etc. A number of instructions which will vary with the type of computer used (about 100) allows operations to be carried out on these data. The programmer is responsible for using the operators with the correct operands. (The computer has no *a priori* knowledge of the meaning of the binary words: it can therefore add letters in among them.) Other instructions allow branches to be formed in the sequence of program instructions. These branches may be:

1. unconditional (the program flow is systematically rerouted);
2. conditional (the rerouting only takes place according to the value of a variable or as the result of an operation);
3. interrupt (an external event or one activated by timing reroutes the program).

This level of programming allows the full potential of the computer to be used, and it allows optimization of the programs in the memory space and from the point of view of the time of execution. On the other hand, this method can be difficult, in particular for large programs. The programs written in machine code are:

1. difficult to write correctly;
2. difficult to correct if errors are found;
3. difficult to reread and, therefore, to maintain.

Since the earliest days of programming, other languages, known as high-level languages have been developed to make program writing easier and more reliable. To be used by a computer, a program written in a high-level language always requires another program which is executed (in machine code) by the same computer or possibly a separate one. Two solutions are possible:

1. *the compiler* translates the program written in high-level language into a machine code, which can then be executed by the computer;
2. *the interpreter* directly executes the program written in high-level language.

The latter solution is most commonly chosen for BASIC and for robotic languages because, despite the lower performance, it is more flexible to apply. The most commonly used languages at present are FORTRAN, COBOL, ALGOL, PL1, APL, Pascal and BASIC (out of the thousands that have been developed). Each of these languages is specific for a certain type of application depending on the type of data offered and the algorithmic structure. It is interesting to note that after the appearance of complex languages (eg ALGOL 68, SIMULA, PL1, APL) in the late 1960s, the 1970s saw the development of far simpler languages (eg Pascal, BASIC). This change in emphasis was not solely due to the reduced potential of the first microcomputers.

The 1980s, however, will almost certainly witness the spread of a new language ADA based on modularity concepts intended to facilitate the writing of complex software and offering real possibilities for programming real-time applications. The other high-level languages mentioned above were not designed with a view to programming applications with real-time control. Additions have therefore been made, with varying degrees of success, to control external processes:

1. definition of external variables in input or output (continuous or Boolean);
2. use of time as an internal variable;
3. definition of different tasks that can be executed in pseudo-parallel form (the processor divides its activity between the different tasks);
4. potential for synchronizing tasks with each other and with the work environment.

Robot languages are generally derived from the real-time versions of the major programming languages (essentially FORTRAN, BASIC, Pascal and ADA) for their general programming facilities, that is on the level of the data structures and associated operations and on the algorithmic level. The following sections give the programming facilities that might be expected from a general purpose language, to provide a basis for comparison with thc facilities offered by existing robotic languages.

4.4.1 DATA STRUCTURES

All high-level languages provide access to the normal interpretations of machine words (usually without the user having to understand the internal representation), and thus allow definition of the data, whether in the form of constants or variables (the data are represented in the program by a symbol, the value of which can be changed before or during the program). The operations that can be carried out on these data will vary in complexity with the language.

Within the framework of robotic languages, it is common for the user not to have access to all these types of data and/or to all the normal operations. Depending on the complexity of the applications to be programmed, this may be a major inconvenience. More serious still is the case of several robotic languages (particularly those inspired by numerical control), in which variables are not permitted and only constants can be used. This removes a considerable part of the advantage provided by the language approach by virtually eliminating the possibility of algorithmic programming (that is not strictly linear).

The different types of data, the operations usually permitted with them and their importance in robotics are described in Sections 4.4.1.1 to 4.4.1.7.

4.4.1.1 Integers

Depending on the language and the storage device used, an integer value is an element of a subset more or less restricted to signed integers. In the majority of minicomputers and microcomputers, in which integers are represented on 16 bits, the scale of values ranges from $-32{,}768$ to $+32{,}767$. In some cases, only positive values are permitted. Some languages and some machines allow the use of integers encoded on 32 bits, ranging from $-2{,}000\,M$ to $+2{,}000\,M-1$ (where $M=1{,}024 \times 1{,}024$). It is possible to express a decimal value as an integer by deciding on a scale factor in advance (generally a power of 10). Thus, the value 53.25 mm could be represented by the integer 5325 with a scale factor of 10^{-2}. If the scale factor remains constant, this type of representation is also called *fixed point*. Normal languages allow the use of the four operations on integers:

+ addition
— subtraction
* multiplication
/ division

and their combination in complex expressions with the use of parentheses. Depending on the type of implementation, the undefined operations (division by zero) or those leading to excess values (outside the limits of the represented scale) are ignored (and give unpredictable results), stop the program or bring about branching into a part of the program intended for the user to process such cases.

All robotic programming languages permit the use of integer values, usually encoded on 16 bits. Depending on the case, this is sometimes limited to constants to specify the coordinates of points or axes, variables also representing coordinates or other values such as row and column pointers for palletization. The use of operators is limited to languages giving access to variables.

4.4.1.2 Real values

In many cases, integers have a scale of values which is too restricted. Real numbers can be used to cover much larger ranges of values because of the use of an integer part with a number of constant, significant figures and an exponent part, multiplying the integer part by a variable factor (a power of ten or two), covering a range from very small to very large numbers (eg from 10^{-24} to 10^{24}). This type of representation is also known as *floating point.* Real values are represented with a limited number of significant figures, which means that the internal representation of a figure supplied by the user is not necessarily exact. The user must, therefore, exercise particular caution during tests on exact values.

The same operations as before are accessible with real numbers, often with exponentiation as well (sometimes limited to integer positive powers). Depending on the languages, the mixture of real and integer operands in a single expression may or may not be allowed. When it is not allowed, functions are generally provided for transforming integers into real numbers (REE X in LM) and conversely (ENT X in LM). Most languages allow the use of normal mathematical functions, such as:

sin	arsin	abs
cos	arcos	exp
tan	artan	log

Few robotic languages offer access to real values, which distinctly limits their potential for acting as general programming languages. Even in robotics, the impossibility of using trigonometric functions (integers and even fixed point numbers do not easily allow this) can be a serious handicap.

4.4.1.3 Boolean values

These are logic data which can have only two values: TRUE or FALSE. They can be represented internally on a single bit or on an entire word (usually 8 or 16 bits). In theory, there are four functions of a single

Boolean variable, given a Boolean result. In practice, only one is used, the complement function:

NOT

In the same way, there are 16 Boolean functions of two logic variables, but only three are generally used:

AND,
OR,
exclusive OR

Various relationship operators in integer and/or real operands also give Boolean values. In general purpose languages, the following operators are usually found:

= equals
> greater than
< less than
>= greater than or equal to
<= less than or equal to
/= unequal

A complex Boolean expression could look like this:

NOT A AND {(X = 1 OR Y = 1) AND (Z > 0)}

Robotic languages only rarely allow the use of Boolean variables. Usually, only the relationship operators in numerical variables are allowed (and generally no potential for logic combination of several tests is provided). This is a handicap in complex applications, such as those in which a number of conditions are to be tested.

4.4.1.4 Character strings

In computer systems, characters are defined by tables which also give their internal representation (on 6, 7 or 8 bits). The most commonly used in the ASCII (American Standard Code for Information Interchange) code table which gives 128 characters on 7 or 8 bits.

A *string* is a sequence of characters, which may be fixed or variable in length (but is frequently limited to 256 characters). Strings are generally used for dialogue with the exterior using the console or via files. They can also be used as identification variables. Operations on strings can often be carried out using general languages. The main ones are:

1. concatenation of two strings;
2. extraction of a substring;
3. length of a string (gives an integer);
4. location of a substring (gives an integer).

Relationship operators for the strings are sometimes available, and give Boolean values:

= equals
< precedence in alphabetical order
> succession in alphabetical order

Robotic languages do not usually allow the use of character strings, except as constants in the dialogue instructions with the operator. This presents a problem when dialogue with files or with another system is necessary.

4.4.1.5 Tables

All general tables allow the use of *indexed variables*, that is variables which can represent all the elements of a table by modifying the indices. These tables may have one dimension (list), two dimensions or n dimensions. In any of these circumstances, the extreme values of each index must be declared in the program (in some languages the lowest value of each index must be 1). Tables are particularly useful in programming when repetitive operations must be carried out on variable data. The variables in a table are usually of the same type: integer, real, Boolean or chain.

indices column → 1 2 3 4

table with two dimensions 5 × 4

	1	2	3	4
1	A(1.1)	A(1.2)	A(1.3)	A(1.4)
2	A(2.1)	A(2.2)	A(2.3)	A(2.4)
3	A(3.1)	A(3.2)	A(3.3)	A(3.4)
4	A(4.1)	A(4.2)	A(4.3)	A(4.4)
5	A(5.1)	A(5.2)	A(5.3)	A(5.4)

indices row ↑

Table 4.2. *Table using indexed variables*

Tables are extremely useful in robotics for representing, for example, lists of points or the many variants of a single object to be assembled. However, tables are only available with the most complex robotic languages.

4.4.1.6 Files

When the amount of data is too large to be declared inside the program (either because of limited memory space or the difficulty of writing them manually), it becomes necessary to store them in an external file.

In a file, the data can be all of the same type (eg point coordinates) or mixed (eg part codes, dates and dimensions all linked to a single object). In this situation, those data which are different but belong to a single class are grouped into records. The exchanges of data between the program and the file are made with the help of the instructions:

READ and WRITE

Depending on the type of data, these instructions may concern a single unit of data, a group of data of the same type or a record. In general, writing and reading are carried out sequentially (*sequential file*). Most high-level languages do, however, permit access to any records inside the file, using a number of access keys (*direct access file* or *random access file*). These files must be recorded on disk, to ensure that delays are as short as possible. Only the most high-level robotic languages allow the use of files (usually only one per program) and access is limited to the sequential mode.

4.4.1.7 Variables and declarations of type

In all textual programming languages, the variables are defined by *identifiers.* Depending on the case, these identifiers may be limited to a single letter, a limited number of letters or an unlimited number. This has a considerable effect on the legibility of the programs. In some languages (known as typified), the programmer must declare the type of each variable before its use (integer, real, Boolean type). In other languages the declaration is implicit and the type of each variable depends on the context in which it is used for the first time. Type languages, although involving some delay, allow verification of the effective use of the variables. Languages known as highly typified such as ADA also allow the range of acceptable values to be defined for each variable, and this range is verified during execution to increase program reliability. The same variety is found in the definition of variables in robotic languages, with the same resulting advantages and disadvantages. The research language AL extends the concept of type right up to the definition of dimension (in the sense of physical meaning) for each variable. The predeclared dimensions are:

time
distance
angle
force
torque
speed
angular velocity
scalar

but other dimensions can be defined by the user. The correct use of variables in the context of dimensional analysis is verified for each expression while the program is being compiled.

4.4.2 THE ALGORITHMIC STRUCTURE

The main purpose of programming languages is the simple and efficient use of the two most important computer instructions:

1. unconditional jumps;
2. conditional jumps (in tests and events).

Unlike other sequential machines (such as numerically controlled first generation machines), computers can endow the machines they control with a certain 'intelligence' with the help of these instructions.

These allow complex programs to be written when they would be too long, if not impossible, to write sequentially, but above all they allow the inclusion of possible reactions to external variables – an essential characteristic of intelligence. Generally speaking, any program whether written in a high-level language or not consists of a linear sequence of instructions. This sequence, however, can be structured to a greater or lesser extent, depending on the language. In structured languages, the instructions are grouped into sections corresponding to a part of the program carrying out a well-defined function. These sections may be:

1. blocks (defined by the instructions START and END);
2. functions;
3. subprograms;
4. whole programs.

In unstructured languages, the branches are made using:

1. labels which are variables or numbers allowing an instruction to be identified;
2. jump instructions of two types:
   – GO TO label *or* JUMP label;
   – IF *Boolean expression* THEN label.

In highly structured languages, jump instructions are not allowed. (These are considered dangerous to use since it is difficult to control their correct use). Instead of jump instructions these languages offer more formalized structures:

1. WHILE structure:
   – WHILE *Boolean expression*
   – DO *instructions block*

2. ITERATE structure:
   - ITERATE *instructions block*
   - UNTIL *Boolean expression*
   (this structure is similar to the previous one, except that the instructions block is always executed at least once)
3. IF structure:
   - IF *Boolean expression*
   - THEN *instructions block*
   - ELSE *instruction block*
4. CASE structure:
   - CASE *Boolean expression 1* → *instructions block 1*
   - CASE *Boolean expression 2* → *instructions block 2*
   ....
   - OTHER CASE *instructions block*
5. FOR structure:
   - FOR *integer variable* = start STEP *integer expression*
   - UNTIL *Boolean expressions* DO *instructions block.*

These structures can be nested within each other: for example, inside an IF expression there may be another IF expression or another conditional structure. In theory, these structures allow the same applications as those programmed by unstructured languages to be programmed. They also involve a certain level of strictness to be attained in programming, which is absent in unstructured languages. This strictness is, however, at the cost of some complexity, and therefore some structured languages also provide access to jump instructions (GO TO label).

Apart from conditional structures, general languages offer the possibility of calling subprograms. A subprogram is a sequence of instructions forming a whole, which can be called at any point in the main program. After execution of a subprogram, the main program continues with its normal progress. The functions are those of the particular subprograms which send back values and can therefore be used inside an expression as a variable. The subprograms and functions are particularly useful when a single sequence of instructions must be used at different points in a single program. They allow the writing and structure to be simplified and the size of programs to be reduced.

Robotic languages offer diverse ways of programming. The simplest offer only very summary instructions for branching without access to the structures or even to the subprograms, whereas others offer the same facilities as the most high-level general languages. The choice again arises between a simple language, but with few sophisticated programming facilities, and a powerful language, which requires an in-depth knowledge of programming techniques.

4.4.3 REAL TIME

With a few exceptions such as LTR (developed for military applications in France) and ADA (developed at the request of the US Department of Defense), all standard programming languages have been developed to resolve problems in which concepts of time synchronization (*real time*) are not involved. These problems have been encountered ever since computers were first used to control machines or processes, and in dealing with them, timing is of prime importance and traditional languages are so far ill-adapted for adequate programming. The difficulty arises essentially from the sequential nature of a program, whereas a control problem is parallel and combinational. This problem is solved by the rapidity of computers in relation to the process to be controlled. The expected response times are generally much more than a millisecond, whereas the execution time of each instruction is of the order of a microsecond. This allows the process to loop very rapidly on a set of relatively independent programs, giving the impression that each program is a permanent active function. Programming these functions is generally achieved by declaring them independent tasks with rules for activation and deactivation, frequency of call and priority. These rules are defined in the main program and are activated by a monitor, a general program put at the disposal of the user, which ensures the management of the time and different tasks. It is this monitor function, absent in general programming languages, which produces the specificity of real-time languages.

The different tasks can be synchronized among themselves by sharing the TIME variable which provides access to the absolute time and to delays between actions, and by making use of the synchronization signals. A task can wait (WAIT FOR *signal* instructions) to continue its progress until a signal is received, emitted by another task (EMIT *signal* instruction).

Robot programming languages are always real-time languages since they are intended to control one or more mechanisms. For reasons of programming simplicity, however, these languages rely very little on the standard real-time programming methods. Often, even the TIME variable is not accessible during programming and in most cases it is not possible to specify the actions which should take place in parallel. Robot languages are usually purely sequential, sometimes with the possibility of starting a new instruction before the end of the preceding one, if it is a displacement instruction.

This limits the possibility of using these languages at present for controlling complex processes, for example, when several machines must be controlled by the same computer. Few languages are capable of controlling several robots cooperatively, that is in parallel.

4.5 Geometric modelling

Because of the nature of the tasks to be undertaken, robot programming languages must exercise perfect control over the movements of robot segments in the work space, or at least over the end effector (gripper or tool). It is possible to restrict the programming to defining the robot's axis motions. This approach, which is valid for programming by training, is extremely difficult in symbolic programming when the positions acquired are not as a result of training but calculation. In high-level robotic languages, the programmer has access to variables which define the positions and orientations of various segments usually in a Cartesian set of coordinate axes. Modelling the position of a segment in a Cartesian set of coordinate axes presents certain problems. The approach generally adopted is to choose a set of coordinate axes associated with the segment and to describe the segment contained in it. The position of the segment is then defined as the positions of this *set of coordinate axes* in the *reference set of axes* (see Figure 4.4).

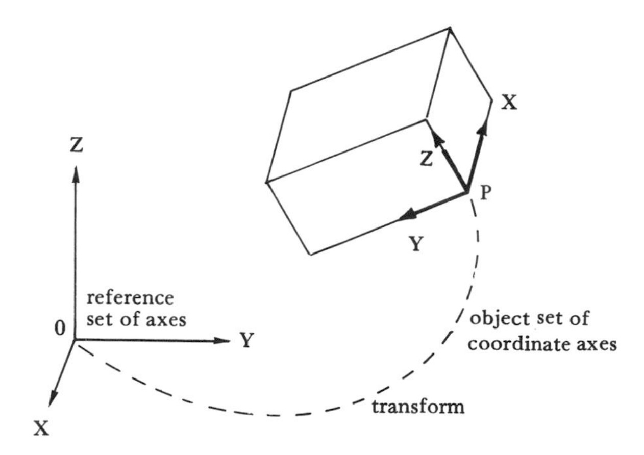

Figure 4.4. *Reference set of axes and segment set of coordinate axes*

The position of one set of coordinate axes in relation to another can be defined using a pure translation between the origin of the reference set of axes and the centre of the segment set of coordinate axes and a pure rotation causing the axes to coincide. Although the translation can be simply defined by the three coordinates of vector 0 P, defining the rotation is more complex. In theory, three variables are sufficient to define this rotation. The Euler angles provide such a representation by three rotations (successively about the Z axis, the new Y axis and the new Z axis).

In practice, the orientation of a set of coordinate axes is often provided to the user by these Euler angles. The position and orientation of the robot end effector will be given on the VDU as:

GRIPPER = (100.0, 100.0, 0.0, 0.0, 90.0, 180.0)

which gives the position according to the gripper set of coordinate axes.

This method of definition using Euler angles does, however, present disadvantages internally, in the calculations for the change in coordinate axis (there are infinite solutions for some degenerated coordinate set changes). Other systems of representation are often used internally. The most commonly used method in robotics is the 3 x 3 matrix, in which each row represents the coordinates of the projections of the new coordinate unit vectors in the old one (see Figure 4.5).

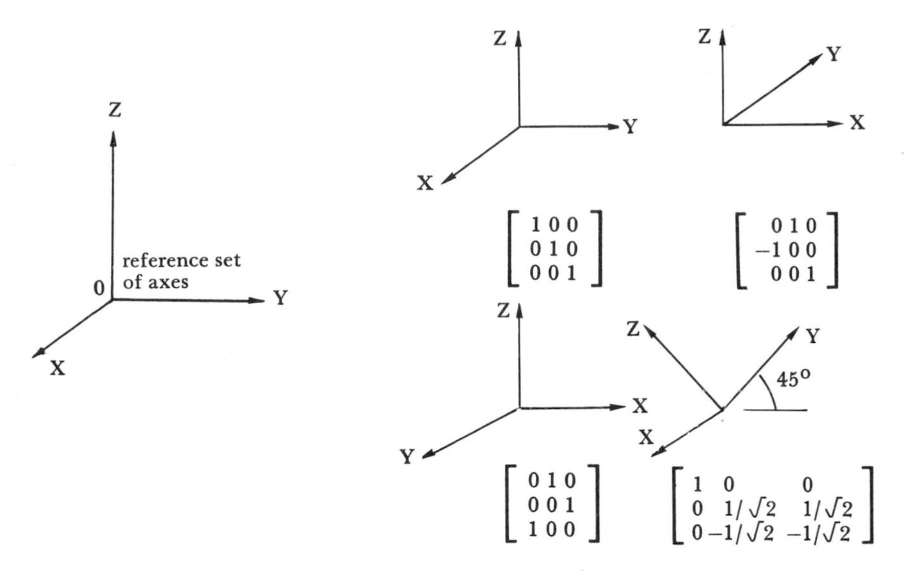

$$\begin{bmatrix} 1 & 0 & 0 \\ 0 & 1 & 0 \\ 0 & 0 & 1 \end{bmatrix} \quad \begin{bmatrix} 0 & 1 & 0 \\ -1 & 0 & 0 \\ 0 & 0 & 1 \end{bmatrix} \quad \begin{bmatrix} 0 & 1 & 0 \\ 0 & 0 & 1 \\ 1 & 0 & 0 \end{bmatrix} \quad \begin{bmatrix} 1 & 0 & 0 \\ 0 & 1/\sqrt{2} & 1/\sqrt{2} \\ 0 & -1/\sqrt{2} & -1/\sqrt{2} \end{bmatrix}$$

Figure 4.5. *Rotation matrices*

Although the level of redundancy is high (nine variables instead of three variables), this method of representation offers a number of advantages:

1. the matrix variables are often used for later calculations;
2. the method avoids the problems of singularity;
3. the representation is relatively easy to visualize.

These 3 x 3 matrices are often enlarged into 4 x 4 matrices, including the translation for forming homogeneous coordinates:

$$M = \begin{bmatrix} \text{rotation} & & X \\ \text{matrix} & & Y \\ & & Z \\ 0\,0\,0 & & 1 \end{bmatrix} \qquad (4\text{-}1)$$

homogeneous coordinate matrix

These matrices can be multiplied together to express sequential coordinate changes, and can be inverted to express the position of the set of coordinate axes:

$$M(R_0 \to R_2) = M(R_0 \to R_1) \times M(R_1 \to R_2) \qquad (4\text{-}2)$$

$$M(R_1 \to R_0) = M^{-1}(R_0 \to R_1) \qquad (4\text{-}3)$$

Some languages (eg AL, LM) make the distinction between the sets of coordinate axes, which are variables describing the location of physical segments, and the transformations, which are operators expressing the spatial relationships between these segments. The coordinate sets and transformations are represented in the same way (translation + rotation) and the difference is purely formal. This distinction will, however, be retained in the following discussion, for reasons of clarity.

A transformation T may be defined either with two sets of coordinate axes R_1 and R_2: it represents the operator bringing about movement from R_1 to R_2:

$$R_2 = R_1 * T \qquad (4\text{-}4)$$

or else directly, by defining its own principal components: the translation and the rotation. These two components are often themselves defined using vectors.

A *vector* is a unit of data with three real components, which allows the location of a point in space, or an oriented segment without fixed origin to be determined. A vector can be defined by its three components X, Y and Z:

$$V = \text{VECT}\,(X, Y, Z) \qquad (4\text{-}5)$$

or by other vectors. The operations generally available for use with vectors are:

— addition of two vectors:

$$V_3 = V_1 + V_2 \qquad (4\text{-}6)$$

— subtraction of two vectors:

$$V_3 = V_1 - V_2 \qquad (4\text{-}7)$$

— change of sign of a vector:

$$V_2 = -V_1 \qquad (4\text{-}8)$$

– multiplication by a real number:

$$V_2 = X*V_1 = V_1*X \tag{4-9}$$

– vectorial product:

$$V_3 = \text{PVECT}(V_1, V_2) \tag{4-10}$$

Other operations on vectors result in real numbers:

– extraction of vector length:

$$X = \text{LENGTH}(V) \tag{4-11}$$

– extraction of components along X, Y and Z:

$$\text{XVECT}(V), \text{YVECT}(V), \text{ZVECT}(V)$$

– scalar product:

$$\text{PSCAL}(V_1, V_2)$$

Three particular vectors are frequently used and may be predeclared; these are the three unity vectors of the fixed set of coordinate axes:

$$V_X = \text{VECT}(1, 0, 0) \tag{4-12}$$

$$V_Y = \text{VECT}(0, 1, 0) \tag{4-13}$$

$$V_Z = \text{VECT}(0, 0, 1) \tag{4-14}$$

From these vectors it is easy to define arbitrary transformation using the two basic transformations.

– Translation along a vector V:

$$T = \text{TRANSLAT}(V) \tag{4-15}$$

is a transformation which allows movement from a set of coordinate axes R_1 to another R_2 by displacing the origin of R_1 along the vector V without any change in orientation (see Figure 4.6).

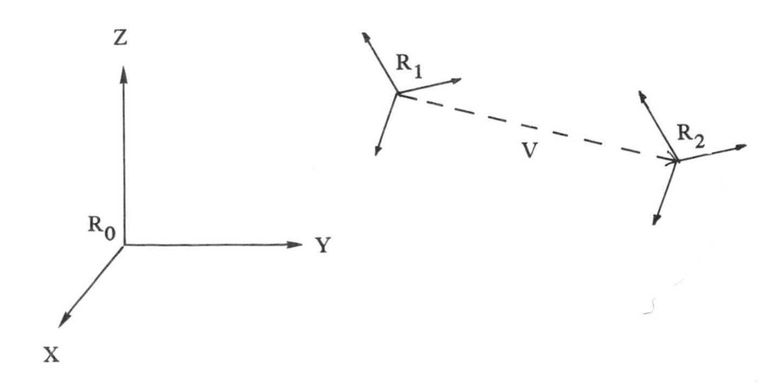

Figure 4.6. *Set of coordinate axes showing a translation*

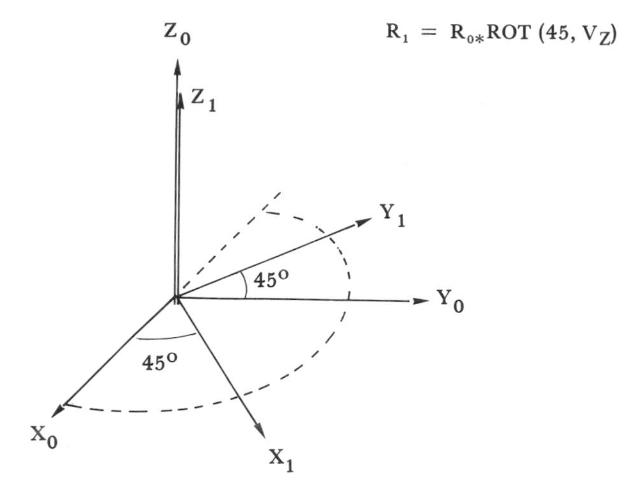

Figure 4.7. *Set of coordinate axes showing a rotation*

– Rotation through angle ALPHA about a vector V:

$$T = ROT\ (ALPHA, V) \qquad \textbf{(4-16)}$$

is a transformation involving the rotation of a set of coordinate axes about an axis V without changing the origin (see Figure 4.7). By simply composing these two basic functions, arbitrary transformations can be made:

$$T = TRANSLAT\ (V_1)*ROT\ (ALPHA, V_2) \qquad \textbf{(4-17)}$$

Depending on the languages the rules of composition can vary. In general, composition is carried out from left to right (LM) or right to left (AL), while applying each new transformation in the newly generated set of coordinate axes, which makes the composition non-commutative. The vectors used in translation and in rotation are usually interpreted in the successive sets of coordinate axes and not as absolute vectors. This is the case in Figure 4.8 assuming that composition is carried out from left to right.

4.5.1 TOOL AND OBJECT SETS OF COORDINATE AXES

In manipulation work, a preferential set of coordinate axes is one which describes the location of the end effector (eg gripper or tool). This set is known as the *tool set of coordinate axes*, generally has the tip of the tool (or the central gripping point of the gripper) as its origin and its axes are oriented as shown in Figure 4.9. In the case of variable tools or grippers (which is the most common case), the tool set of coordinate axes is deduced from an intermediate set, called the *wrist set of coordinate axes*, using transformation which depends on the tool used and can be modified by the user.

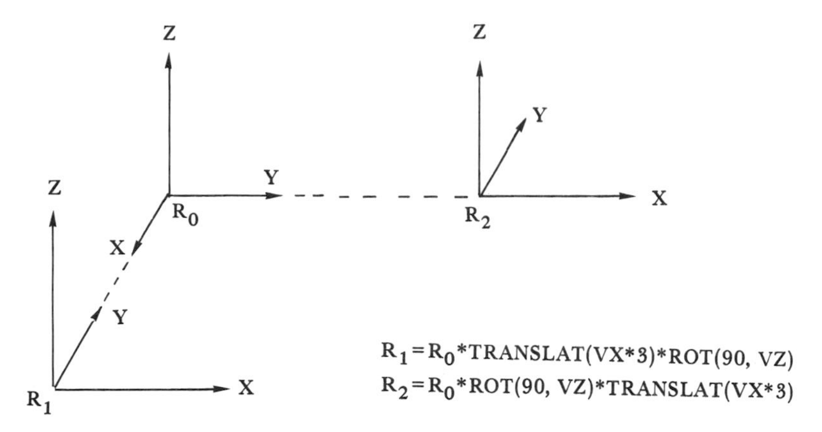

Figure 4.8. *Multiplication of transformations*

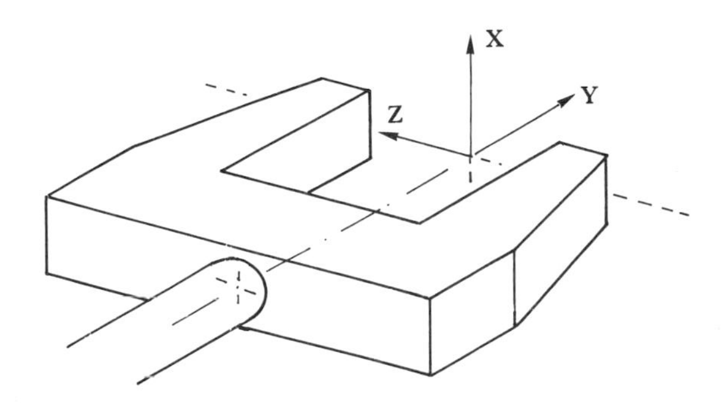

Figure 4.9. *Tool set of coordinate axes*

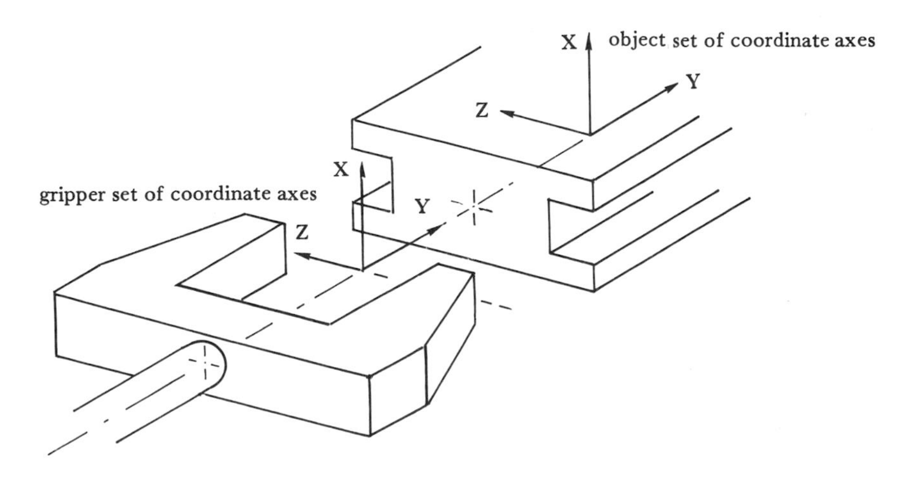

Figure 4.10. *Gripping an object*

Robot programming is essentially making the tool set of coordinate axes coincide with the others linked to the task to be executed. If an object is to be manipulated, the user must define one or more gripping sets of coordinate axes for each object, bearing in mind the characteristics of the gripper, so as to make its set of coordinate axes coincide with that calculated for gripping motions (this involves the geometric dimensions and the direction of approach) (see Figure 4.10).

4.5.2 CONNECTIONS BETWEEN SETS OF COORDINATE AXES

Some robot languages (eg AL, LM) allow automatic updating of the sets of coordinate axes when the objects are moved. Some sets of coordinate axes can be linked to a single object, whether temporarily or permanently: for example, the gripper and the object it grasps, two assembled objects or else two sets of coordinate axes for gripping a single object. Using a linking instruction (ATTACH IN LM), several sets of coordinate axes can be moved simultaneously. The inverse instruction (DETACH) allows the sets of coordinate axes to be made independent of one another.

4.6 Movements

Generally speaking, in the end effector level programming languages currently in use, programming of robot motions is carried out in the point-to-point mode, with the destination of the tool set of coordinate axes specified (or of a set of coordinate axes linked to the tool):

MOVE GRIPPER TO OBJECT

or by specifying relative movements in relation to the current position:

MOVE GRIPPER BY ΔX, ΔY, ΔZ, $\Delta\alpha$, $\Delta\beta$, $\Delta\gamma$.

Depending on the language, these two basic instructions may or may not accept the variants described in Sections 4.6.1 to 4.6.5.

4.6.1 STRUCTURE OF THE TRAJECTORY

The trajectory may be executed in the free mode (trajectory with a low level of precision), the coordinated actuator mode (the actuators follow congruent speed laws) or linear interpolation in the Cartesian mode (the origin of the tool set of coordinate axes follows a straight or circular path). Depending on the language and the robot, the user either has free choice of these three modes, to choose between two or, in some cases, no choice at all.

Example: MOVE GRIPPER TO OBJECT IN CARTESIAN MODE

4.6.2 SPEED OR TIME OF EXECUTION

In many applications, it is desirable to perform the trajectory either at a given speed (eg welding, deburring, applying glue) or in a given time. Most current languages do not provide access to these two essential parameters because of the complexity of calculation involved. On the other hand, it is often possible to provide a speed factor relative to the nominal speed.

Example: MOVE GRIPPER TO OBJECT AT SPEED = 0.25

4.6.3 INTERMEDIATE POINTS

In complex trajectories, or if it becomes necessary to avoid obstacles, it is useful to be able to specify a trajectory made up of several segments (in the free, coordinated or linear mode) without the robot stopping at each intermediate point. This may be achieved either by specifying intermediate sets of coordinate axes inside a movement instruction:

MOVE GRIPPER VIA R_1, R_2, R_3 TO OBJECT

or by linking several movement instructions:

CONTINUOUS
MOVE GRIPPER TO R_1
MOVE GRIPPER TO R_2
MOVE GRIPPER TO R_3
MOVE GRIPPER TO OBJECT
END

In general, these intermediate sets of coordinate axes are not reached with great precision: the tool passes close to them. It may be useful to specify the maximal value for the acceptable gaps in each variable. As this is not convenient for all the variables in the set of coordinate axes, it may be simpler to specify only the radius of a sphere through which the end point must pass. Unfortunately, not even this option is available for languages currently in use.

4.6.4 APPROACH AND DEPARTURE POINTS

In robotics it is frequently useful to have to approach the final position (eg, to grip an object) with a final phase which is not necessarily in the direct trajectory arising from the preceding point. It is, therefore, generally necessary to specify an intermediate point close to the end point. A simplified version of this approach is to specify only an approach distance: the intermediate point is generated automatically as a set of coordinate axes with the same orientation as the destination

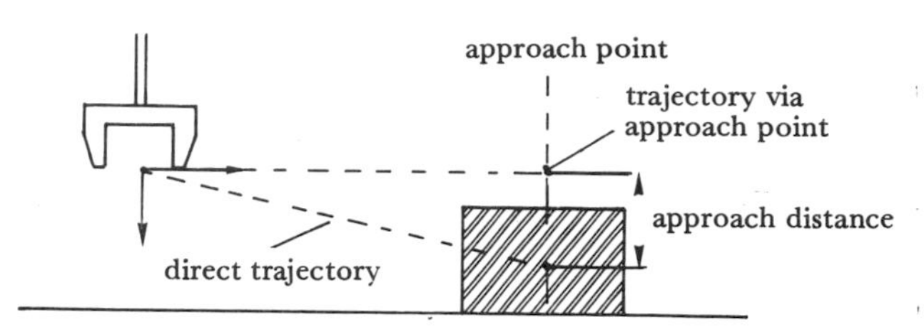

Figure 4.11. *Approach point*

set of coordinate axes, and translated along the tool axis by a specified distance (see Figure 4.11).

Example: MOVE GRIPPER TO OBJECT WITH APPROACH = 50

A similar specification can be made for the start of a trajectory with a starting distance.

4.6.5 COMPLEX TRAJECTORY

Sometimes trajectories more complex than straight lines or circular paths are necessary in certain types of application. The two possible approaches that can be used to generate trajectories of this type follow.

A generative approach in which the trajectory is generated in the point-to-point mode at the maximum sampling frequency or else by modification of the speed vector.

Example: MOVE GRIPPER CONTINUOUSLY BY D_X, D_Y, D_Z,
D_A, D_B, D_C WITH
D_X = arithmetical expression
D_Y = arithmetical expression
....
END

An approach involving training in which the trajectory has been recorded, for example, in manual mode. In this situation the trajectory is represented by a complex variable made up of successive positions in the Cartesian space. In some cases, transformations can be carried out on this type of trajectory.

Example: MOVE TOOL ALONG TRAJEC-A

None of the languages currently available allows generation of such complex trajectories (in the generative or training mode).

4.7 Sensors

All robotic systems use sensors to make their actions take account of their environment. In the simplest and by far most common instances, these are on-off devices testing say, for the presence or absence of a piece part, or the operation of a nearby machine. In some cases it might be useful to have access through programs to continuous variables. The velocity of a conveyor belt is one such example. Unfortunately many languages do not make provision for such variables. Software can make use of continuous or binary variables to initiate branching sequences in a program or to terminate an action. The detection of a contact with an object, or an unexpected force could be such a case.

Example: MOVE TOOL TO DEST UNTIL $F_Z > 10$

It is desirable, but rarely possible, to modify a movement once it has started using servocontrol of external variables (eg of forces or over distances from a target as in seam tracking in welding robots). For this to be possible, it is necessary to have access to the movements of the articulations in the reference set of coordinate axes both for position and velocity.

Example: MOVE TOOL WITH $V_X = 50$ and $V_Y = (A - B)*5$

where A and B would be two differential values supplied by a proximity sensor.

Other sensors specific to robotics providing more detailed information have been developed and used with the languages available. These are strain gauges or sensors with three to six components and vision sensors. A strain gauge could be used to define, at language level, a rigidity matrix which would allow the robot to simulate a spring effect supporting the tool. This rigidity matrix is defined by three axial rigidities (along axes X, Y and Z) (eg expressed in N/m) and three rotational rigidities (about the three axes) (eg expressed in Nm/radian). The movement would be specified along a predetermined trajectory but variations in the trajectory would be accepted as a function of the specified rigidity.

Example: MOVE OBJECT TO HOLE WITH RIGIDITY = (10, 10, 100, 10, 10, 10)

for a vertical insertion with high rigidity along the Z axis and a high level of compliance in the other axes and rotations.

Vision sensors employ an external process to examine a scene and send, on demand from the main program, a certain number of variables which depend on the vision system. In its simplest form the vision system sends an object number corresponding to a recognized shape and the position and orientation of the object. Many systems of this type are available handling only flat objects in a plane with a single part in the field of vision. More complex experimental systems allow simultaneous recognition of several objects with or without overlapping in a plane, extraction of significant parameters relating to an object (eg perimeter, shape factor, number of holes, surface area of object and of holes). The program must utilize the data provided by the vision system in order to control the robot according to the programmer's wishes. The robots on the market can only use information from their own visual sensors and not from vision systems made by other manufacturers.

4.8 Tools

In addition to carrying out displacements, a robot must be able to operate tools. These may be situated at the end of the arm (eg gripper, torch, spray-gun, buff-wheel) or at a fixed post (eg buff-wheel, sander, vice, pivoting table, welding gripper). Generally speaking, tools can be classified into two types: those controlled in the binary mode and those controlled in the continuous path mode. Depending on the languages and the tools, standard outputs control commands (binary or continuous path) or else commands specific to the tool can be used.

Example: ACTUATE TOOL 1 (A, B)

where A and B are operating parameters of TOOL 1 (eg a voltage and a wire speed for a welding torch).

One specific tool to be considered is the proportional opening gripper. It is not only possible to give a closing or opening command with a required gap but it is also possible to check that the correct gap has been obtained (eg to verify that an object has been grasped correctly).

Example: CLOSE GRIPPER TO 12.5 IF NOT GO TO WITHOUT-OBJECT

4.9 Multi-robot operations

Some programming languages allow several robots to be controlled at once. Unfortunately, the performance standards currently obtained from control systems prohibit the simultaneous control of a large number of axes if calculations for the transformation of coordinates are involved. Control is therefore carried out sequentially (one robot moves, then another) or else with simple structures (eg Cartesian robots).

Nevertheless, the possibility for simultaneous control exists and will become increasingly real with the improvement in performance of the control structures. The difficulty lies in the programming of simultaneous motions. Two techniques are currently available.

1. *Sequential programming* in which a new instruction can be activated without waiting for the end of the preceding one (in the case of movements).

   Example:
   ```
   MOVE ROBOT 1 TO POS 1
   WITHOUT WAITING
   MOVE ROBOT 2 TO POS 2
   ```

   In this example, the two robots ROBOT 1 and ROBOT 2 will start their individual movements almost simultaneously.
2. *Parallel programming* with two programs which appear to be carried out simultaneously:

   ```
   IN PARALLEL
   START 1
      MOVE ROBOT 1 to POS 1
      ......................
      ......................
      END 1
   START 2
      MOVE ROBOT 2 to POS 2
      ......................
      ......................
      END 2
   ```

The second approach is the more general, and is very similar to real-time language structures.

4.10 Some commercial languages

This section contains a presentation of the major programming languages available on the market. This list is not exhaustive, and many further languages will appear in the months to come. The selection criterion has mainly been the availability of reference manuals.

4.10.1 AL

AL (Arm Language) is currently the language which has undergone the most important development. It first appeared in the course of original research at Stanford University into programming robots and the use of WAVE language, and has been the object of constant improvements since 1974. In its first version, it was broken down into a planning module, compiling programs on a PDP-10 and generating trajectories. An intermediate code was then executed on the execution module, using PDP-11/45. POINTY language, similar to AL was developed from 1975 onwards on a PDP-11 so as to offer a more interactive version in terms of programming.

A new interactive version of AL was introduced recently, almost entirely in Pascal OMSI on a PDP-11/45. This new version, which is particularly useful for controlling Puma robots, can be obtained commercially. AL has been taken up outside the USA, in particular, two versions of AL have been introduced at the University of Karlsruhe and at the University of Tokyo.

4.10.2 AML

AML (A Manufacturing Language) is the language represented by IBM with its robots, and was introduced onto the market in 1982: the Cartesian hydraulic robot RS-1 (or 7565) produced by IBM and the 7535 electric robot from the Scara range, produced in Japan by Sankyo. The language and the Cartesian robot are the results of important research work carried out in the laboratory at Yorktown Heights in the mid 1970s. A series of languages was produced there, and applied in assembly experiments (MAPLE, EMILY and AUTOPASS), and a number of the robots were used in actual production by IBM in its Florida factory from 1979 onwards. With its top of the range robot RS-1 (controlled using a Series 1 minicomputer), IBM was the first manufacturer to produce a standard version system equipped with strain gauges in the fingers of the gripper. The AML version, which comes with the Sankyo robot is a subset of AML: the programs are developed on an IBM-PC microcomputer and stored in the standard control system produced by Sankyo.

4.10.3 IRL

IRL (Intuitive Robot Language) is a language developed by the Swiss company Microbo, affiliated to the large watchmaking companies, for its range of high precision assembly robots. This range is made up of the electric model Souris with six DOF (precision 1/100 mm), the hydraulic model Castor (in the process of being replaced by a similar electric model) and the Ecureuil model introduced in 1983. These robots have been used exclusively in the watchmaking industry in Switzerland for the last three years, but Microbo decided to make them generally available on the market during 1983. The research and development work for this range arose partly from the research work carried out at the Institut Polytechnique de Lausanne.

4.10.4 LM

LM (Langage de Manipulation) was developed in the IMAG robotics laboratory at the University of Grenoble in 1979. It owes much of its inspiration to AL, adopting most of its main concepts (apart from parallel programming), but it is used on a microcomputer (LSI-11/23 and 68000). An agreement was signed with Itmi, which was founded on the initiative of a number of laboratory researchers in robotics. Itmi markets LM language which may thus be acquired by other robot producers who do not wish to invest in the development of control methods. At the time of writing LM has been licensed to Scemi, Matra and Hewlett Packard.

4.10.5 LPR

LPR (Langage de Programmation de Robot) was first developed in 1978 for the control of ACMA robots, produced by Renault (also called Version 5 control or V5). The preceding control system (or V3) used minicomputers (Mitra-105 or Philips P-850) programmed in assembler in a specific manner for each class of application (basically point-to-point). To respond to more complex internal requirements, a new processing structure based on several 16-bit microprocessors and on bit slice processors was developed. To facilitate the development of complex applications, LPR was designed using a Grafcet approach. In this approach, the development is done on powerful machines (VAX, Mitra or Philips minicomputers) and the code is down-loaded to the microprocessors. At this time, LPR is still considered by Renault as a tool for internal development to which the users would not have access.

4.10.6 MCL

MCL was developed within the framework of the ICAM projects (Integrated Computer-Aided Manufacturing, financed by the US Department of Defense) to resolve in a unified way all the problems associated with robot programming. The project was entrusted to McAuto (affiliated to McDonell Douglas) which now offers a commercial interactive version (MCL/11) using a PDP-11 (the original ICAM version used an IBM-370 in batch mode). MCL is an extension of APT and its aim is the programming of flexible units, that is of a set of machines served by one or more robots. At the present time, MCL can control Cincinnati T3 and Westinghouse Allegro robots.

4.10.7 PLAW

PLAW (Programming Language for Arc Welding) is included because it demonstrates the way in which textual languages may one day supplant training techniques. PLAW is particularly well-suited to ‘intelligent’ welding, that is, welding which involves the use of sensors (eg seam tracking). It was developed by Komatsu for its series RW Cartesian robots, equipped with arc current sensors and television cameras.

4.10.8 RAIL

RAIL was developed by Automatix (formed in 1978 by the founder of Computervision and a number of researchers from MIT and Stanford including V. Scheinman formerly of Unimation and Vicarm), and is the first robot language that can be applied to problems of manipulation as well as problems of vision. It is largely based on Pascal, and is programmed and executed on a machine developed by Automatix using a 6800 microprocessor. The robotic applications mainly relate to continuous path control welding (on a Hitachi robot) with the possibility of using sensors for seam tracking and assembly (on a DEA robot).

4.10.9 ROL

ROL (RObot Language) is based on the new approach to the robotics market, similar to Itmi. ROL was also developed from research (with the LAMA-S language developed during the Spartacus project between 1976 and 1979) with the aim of designing a complete commercial system for computerized control (including hardware and software) adaptable to any robot. The novelty of this comes from the fact that GIXI (a software house affiliated to French CEA) does not have a policy of selling robots itself. Its market, therefore, consists of manufacturers of robots (and technically similar machines) and engineering firms. The system made available on the market by GIXI in 1983 is highly modular and can be used for very simple systems (eg of two axes) as well as for configurations involving several robots, processors and sophisticated sensors. Programming is carried out on an IBM-PC microcomputer.

4.10.10 SERF

SERF (Sankyo Easy Robot Formula) was developed in 1978 by Sankyo for its assembly robot. It was developed in the course of university research for the SCARA project, directed by Professor Makino. The Sankyo robot is oriented towards small-scale assembly tasks in which palletization plays an important role, and this textual language was the first introduced into Japan. The language functions on a simple level and in the style of numerical control languages, nevertheless, it permits the use of loops and tests. The development of the application program is carried out using a specialized console (on-line or off-line) which may be abandoned once the program is complete. Control is based on a single Z-80 microprocessor.

4.10.11 SIGLA

SIGLA (SIGma LAnguage) was the first commercial language available for use with an industrial robot. It was developed in 1974 by Olivetti for its Cartesian Sigma robots, and was considerably influenced by numerical control languages, but nonetheless allows control of several arms, with loops and tests on the sensors. The grippers, in particular, may be equipped with strain gauges which allow assembly operations to be monitored. Sigma robots are intended for three main applications: assembly, drilling (or small-scale machining) and welding. Sigma robots are now manufactured in the USA by Westinghouse in parallel with the Puma robots. It seems likely, therefore, that SIGLA will enter into competition with VAL.

4.10.12 VAL

VAL (VicArm Language) was developed in 1973 by students at Stanford University. It is based on WAVE language and used with the small research robots retailed by Vicarm (Victor Scheinman). Several of these robots were sold to large industrial groups and were used for their robotic development projects. In 1977, Vicarm was bought by Unimation in order to develop the Puma programme (commissioned by General Motors), which gave rise to a range of Puma robots using VAL as their programming language. Recently, VAL and its associated hardware have been adapted to all the Unimate robots which previously used an old form of point-to-point control. A new version of VAL, called VAL II, is at present being tested. It should offer a number of improvements over VAL based on the same concepts.

4.11 Examples of programming

As in the simple instructions examples presented in previous sections, an invented but similar type of language adopting some of the most advanced concepts in current use is used here. The examples demonstrate a programming principle and have been deliberately simplified to allow them to be easily understood. Real programs would be more complex, and would include many more tests to ensure safe conduct of the manipulation.

The program is written in capital letters. The explanatory comments are written in lower case letters and follow the instruction they qualify.

Example 1: Machine loading and unloading (see Figure 4.12)

READ GRASP, MACHINE, MACHINE-APPRO, DEPOSIT, WAIT

all these variables are for the destination or transit point coordinate sets.

language	robot(s)	developed by	dates	execution computer	programming computer	language type
AL	Vicarm Puma	Stanford University	1974–1983	PDP 11/45 + n6502	PDP 10 or PDP 11	ALGOL
AML	IBM Sankyo	IBM	1977–1982	Series-1 Z-80	same*	APL
IRL	Castor Souris	Microbo	1978–1980	n × 8085	same	NC
LM	Scemi +	University of Grenoble ITMI	1979–1983	LSI-11/23 + n68000	same	Pascal
LPR	ACMA (all)	RNUR-DTAA	1978–1983	2 × 8086 + AMD	VAX, Mitra Philips or Texas	Grafcet + machine code
MCL	Cincinnati	McAuto	1979–1982	NC	PDP-11	NC
PLAW	Komatsu-RW	Komatsu	1980–1982	NC	PDP-11	NC
RAIL	Hitachi DEA	Automatix	1979–1982	68000 + ?	same	Pascal
ROL	any	Gixi	1979–1983	n × 8086 + 8087	IBM-PC	BASIC
SERF	Sankyo	Sankyo	1977–1979	Z-80	Z-80	NC
SIGLA	Sigma	Olivetti	1974–1976	NC	same	NC
VAL	Puma (all) Unimate (all)	Vicarm Unimation	1974–1978	LSI-11 + 7 × 6502	same	BASIC
VAL II	Puma (all) Unimate (all)	Unimation	1979–1983	LSI-11/23 + 7 × 6502	same	Pascal

* Same means that the programming computer is the same as the execution computer.

Table 4.3. *General characteristics of the major programming languages*

language	integer values	real values	Boolean values	chains	tables	declaration of type	operations arithmetic	operations Boolean	mathematical functions
AL	no	variables	variables + relationships	variables	n homogeneous dimensions	yes + dimension	yes	yes	variables + relationships
AML	variables	variables	relationships	2 dimensions	yes heterogeneous dimensions	yes	yes	relationships	
IRL	variables	no	relationships	constants	no	no	yes	no	relationships
LM	variables	variables		variables	2 homogeneous dimensions	yes	yes	yes	
LPR	variables	variables	relationships	variables	3 dimensions	no	yes without parentheses	yes	relationships
MCL	variables	variables	yes	variables	yes	no	yes	yes	yes
PLAW	constants	no	signals	no	no	no	no	no	signals
RAIL	variables	variables	expressions	variables	2 homogeneous dimensions	no	yes	yes	expressions
ROL	variables	variables	expressions	variables	no	yes	yes	yes	expressions
SERF	variables	no	signals	no	no	no	yes	no	signals
SIGLA	variables	no	signals	constants	no	no	yes	no	signals
VAL	variables	no	expressions	constants	no	no	yes	yes	expressions
VAL II	variables	no	variables	constants	yes	no	yes	yes	variables

Table 4.4. *Types of data*

language	structured programming	labels	parallelism	interruption of movement	synchronization (signal-wait)	subprogram function	access to time
AL	yes	no	yes	no	yes	yes	yes
AML	yes	no	between movements	yes	no	yes	no
IRL	no	yes	no	yes	no	yes	no
LM	yes	yes	between movements	no	yes	yes	yes
LPR	no	yes	yes	yes	yes	yes	no
MCL	no	yes	no	yes	no	no	no
PLAW	no	yes	no	no	yes	yes	no
RAIL	yes	no	no	no	yes	yes	no
ROL	yes	yes	yes	yes	yes	yes	yes
SERF	no	yes	no	no	yes	no	no
SIGLA	no	yes	no	yes	yes	yes	no
VAL	no	yes	no	yes	yes	yes	no
VAL II	yes	yes	yes	yes	yes	no	no

Table 4.5. *Algorithmic structure*

language	transformation of sets	attachment of sets	trajectory	intermediate points	approach/departure	velocity	compliance
AL	yes	yes	coordinated	yes	yes	percentage	yes
AML	no	no	coordinated	no	no	percentage	no
IRL	no	no	coordinated	no	no	percentage	no
LM	yes	yes	coordinated/ linear	yes	yes	percentage	no
LPR	no	no	coordinated/ linear	yes	no	absolute	no
MCL	no	no	coordinated/ linear/ circular	yes	no	percentage	no
PLAW	no	no	coordinated/ circular	yes	no	percentage	no
RAIL	yes	no	coordinated/ circular	yes	yes	percentage	no
ROL	yes	yes	coordinated/ linear/ circular	yes	yes	absolute	no
SERF	no	no	non-coordinated	no	no	percentage	no
SIGLA	no	no	coordinated	no	no	percentage	no
VAL	yes	no	coordinated/ linear	yes	yes	percentage	no
VAL II	yes	no	coordinated/ linear	yes	yes	absolute	no

Table 4.6. *Movements*

language	manual training	program storage	networking	step by step	vision system	stress sensors
AL	yes (new version)	PDP-11	yes	no	yes	yes
AML	yes	series-1	yes	yes	no	yes
IRL	yes	diskette	no	yes	no	no
LM	yes	diskette	no	no	yes	yes
LPR	yes	diskette	yes	no	yes	yes
MCL	no	PDP-11	yes	no	yes	no
PLAW	yes	cassette	no	yes	yes	no
RAIL	yes	diskette	yes	yes	yes	yes
ROL	yes	diskette	yes	yes	yes	yes
SERF	yes	cassette	no	yes	no	no
SIGLA	no	ruban type	yes	yes	no	yes
VAL	yes	diskette	no	yes	yes	no
VAL II	yes	diskette	yes	yes	yes	yes

Table 4.7. *Programming and extensions*

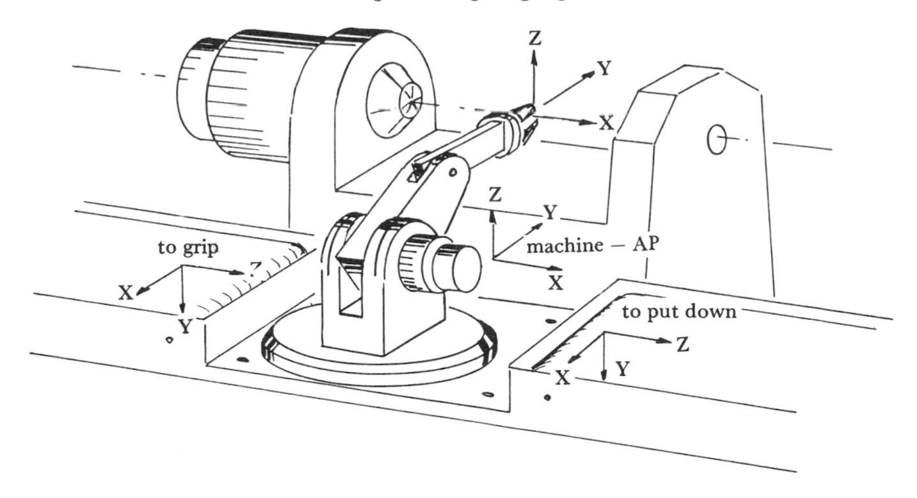

Figure 4.12. *Loading-unloading operation*

MOVE GRIPPER TO WAIT – OPEN GRIPPER
WHILE GO = TRUE DO

GO is an input variable describing the state of the station control switch.

MOVE GRIPPER TO GRASP WITH APPROACH = 10
IF PART-OK THEN CLOSE GRIPPER
ELSE INCIDENT

PART-OK is a switch which indicates the presence of a part at the grasping station. INCIDENT is a function (not described here) which calls the operator.

MOVE GRIPPER TO MACHINE VIA MACHINE-APPRO

MACHINE-APPRO is an approach set of coordinate axes for loading the machine.

SIGNAL CLOSE-JAWS
(sending a command to close the jaws of the lathe)
OPEN GRIPPER
MOVE GRIPPER TO WAIT VIA MACHINE-APPRO
SIGNAL START-MACHINE
(activation of the lathe cycle)
WAIT END-MACHINE
(wait for command to end cycle)
MOVE GRIPPER TO MACHINE VIA MACHINE-APPRO
CLOSE GRIPPER
WAIT 2 SIGNAL OPEN-JAWS
MOVE GRIPPER TO DEPOSIT VIA MACHINE-APPRO WITH

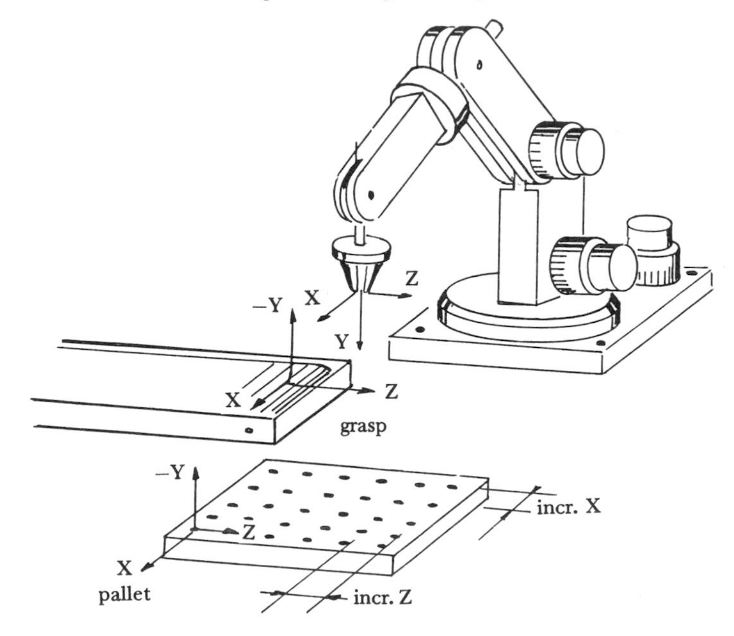

Figure 4.13. *Palletization*

```
APPROACH = 20
(finished piece deposited)
OPEN GRIPPER
MOVE GRIPPER TO WAIT WITH DEPART = 20
(return to intermediate position and recommencement of cycle)
END
```

Example 2: Palletization (see Figure 4.13)

```
READ GRASP, PALLET, INCR-X, INCR-Z, NX, NZ
```

GRASP is the set of coordinate axes for the arrival of the objects, PALLET is that of the first hole in the pallet, INCR-X and INCR-Z are the distances between the rows and columns and NX and NZ are the number of rows and columns.

```
I = 1, J = 1
DEPOSIT = PALLET
OPEN GRIPPER
MOVE GRIPPER TO GRASP WITH APPROACH = 10
  WHILE I ≤ NX DO
     WHILE J ≤ NZ DO
     CLOSE GRIPPER
     MOVE GRIPPER TO DEPOSIT WITH DEPART = 10,
     APPROACH = 10
     OPEN GRIPPER
```

```
    MOVE GRIPPER TO GRASP WITH DEPART = 10,
    APPROACH = 10
    DEPOSIT = DEPOSIT* TRANSLAT (INCR-X, 0, 0)
    update for following deposit (following column)
    J = J + 1
    END
DEPOSIT = PALLET* TRANSLAT (0, INCR-Z* I, 0)
movement to column 1, row I.
I = I + 1
END
```

Chapter 5

CAD robot programming

5.1 Introduction

The research carried out in recent years into the use of computer-aided design (CAD) for robotic applications and the various industrial applications which have been developed recently in this field relate to different imperatives and requirements. Programming, or more generally using, robots often involves problems related to the movement of an articulated structure in a work space in which there are obstacles present, with a view to carrying out a given task. This type of problem, which requires the user to choose the best robot installation for the job, is often too complex to be solved by the buyer alone. It is at this level that the first major advantage of using CAD, or rather computer graphics, appears. CAD also allows:

1. development to be anticipated;
2. development to be carried out using a powerful computer;
3. time and costs of development to be reduced;
4. risks during programming to be minimized;
5. dependence on the production of a prototype to be avoided.

As for the programming itself, the increase in productivity brought about by the use of an off-line programming method is considerable. As seen in Chapter 3 in connection with servocontrolled robots, the architecture of a robot can allow many automated tasks to be programmed. Whether these tasks involve, for example, spot welding or paint spraying with seam tracking, the fact remains that the programming required often consists of a training method. As a result, the need to reprogram a single robot may immobilize an automated process in which it is integrated. This aspect of on-line programming is particularly inconvenient when the robot is used for small- or medium-scale production. It may also limit the flexibility of automated workshops.

Current research and industrial applications are attempting to use CAD systems for off-line programming or, at least, to minimize the involvement of the robot itself in its reprogramming. CAD systems are already widely used in industrial processes. This ranges from the installation of robots in workshops or factories and the design of mechanical parts to the development of programs for assimilation of numerically controlled machines and the scheduling of work loads. If CAD is to improve productivity in the future, by allowing robots to be programmed off-line, its use must be considered in the more general framework of design and production techniques, in order to allow its importance in robotics to be assessed. At present, CAD provides answers to the problems raised by the use of robots in industry. Such diverse automated tasks as paint spraying, welding, materials handling and even assembly, can benefit from off-line CAD.

The aim of this study is to analyse the advantages of CAD in the execution of these tasks and how programming languages or sensors, for example, might be improved to allow a more effective approach to robotics in the context of automated design and production systems. Applying CAD to robots in a flexible workshop, in which it would already be applicable to numerically controlled machines on an industrial level, seems promising for the future.

5.2 General description of a CAD system

This chapter describes the CAD system in terms of its architecture, hardware and software. This is intended to help the way in which CAD is used for the definition and visualization of an object (see Section 5.2.4).

5.2.1 ARCHITECTURE

A CAD system is based on a standard processing system comprising (see Figure 5.1):

1. a central processing unit;
2. one or more work disks with large capacity memory (RAM);
3. a device for reading magnetic tapes, which permits the use and ensures the storage of the system utilities, programs and graphics files designed by the user;
4. peripherals, such as alphanumeric consoles and printers.

A number of peripheral devices can be added to this processing architecture;

1. graphics inputs, which allow the X, Y coordinates of points in a plane to be rapidly sent to the computer: digitizing tablets, light pen, mouse, track ball etc;
2. visual display units (VDUs), which, in conjunction with the graphics input devices, allow a geometric data base to be constructed for

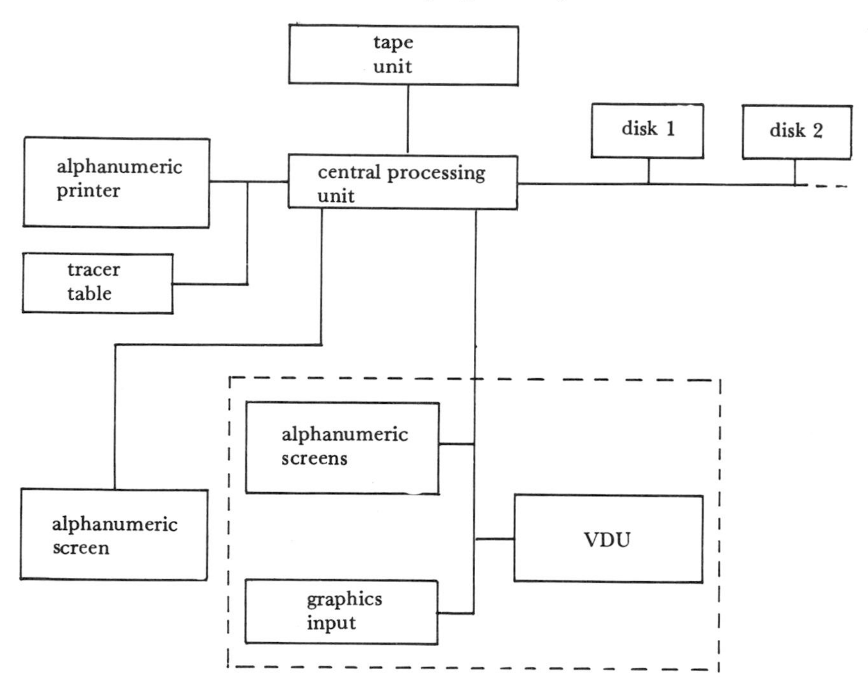

Figure 5.1. *Design of a graphics workstation*

any object defined in two or three dimensions, and to use the data base by displaying the object on the VDU;

3. various types of tracing tables (with rollers, drums, electrostatic devices etc) which can be used to produce plans of any object defined in the data base.

5.2.2 HARDWARE

The CAD system is generally designed to be used with a number of workstations for reasons of cost-effectiveness: a system with four graphics workstations, four alphanumeric workstations and a drum tracer costs approximately $80,000. Several workstations, based on alphanumeric consoles, can be included, allowing the user to design and perfect any type of program (making use, if appropriate, of the purely graphics facilities of the system), as well as a number of graphics workstations. The visual display of images on a graphics workstation can be achieved in a number of ways. Scanning tubes (raster or vector type) and memory tubes are the most widely used screen technologies (see Table 5.1).

	definition (max)	advantages	disadvantages
scanning tubes		easy interactive usage possibility of generating colour images	flickering
1. television	1640 x 1320	possibility of generating realistic images	aliasing
2. calligraphic	2048 x 2048	lower cost	
3. memory tubes	4096 x 4096	high definition lower cost	must erase to modify

Table 5.1. *Advantages and disadvantages of VDUs*

1. *Television (raster) type scanning tubes*: the image is formed by modulation of the electron beam, which continually scans the entire screen, making up a frame of horizontal lines.
2. *Vector type scanning tubes*: the process is similar to that used with oscilloscopes; the beam is subject to x and y servocontrol to draw the lines to be displayed.
3. *Memory tubes*: the memorization of the signal is carried out by the maintenance of secondary emission, which is produced by the impact of the beam on the screen.

In 1 and 2 it is necessary to refresh the image 25 or 50 times per second (depending on the type of terminal used) and, therefore, to store the image in semiconductor memories. Figure 5.2 shows a typical graphics workstation.

The generation of colour images is possible using a scanning tube; this technique can reduce the risks of interpretative errors and allows man-machine dialogue. The generation of 'solid' images (ie that look realistic) is possible only on VDUs with raster scanning. The importance of the size of the VDU must be stressed, since it must be sufficiently large to allow efficient use with multiple views.

The system itself is constructed around a computer, minicomputer or multiprocessor architecture which works with 16- or 32-bit words. In all systems at the top of the range, one or more graphics processors manage the systems graphics inputs and outputs. At the level of the graphics workstation, trends towards 'local intelligence' in this type of terminal should be noted: a specialized microcomputer can be used to relieve the central processing unit of a number of functions, such as:

Figure 5.2. *Graphics workstation (from Peugeot-Citroen)*

1. generation of circles, straight lines, ellipses etc;
2. calculation of transformations (translations and rotations) required for image generation.

As a result of this, the criterion for the choice of VDU must take into account the facilities offered by the workstation software.

5.2.3 SOFTWARE

Unlike the basic software used in any standard system, the software for a CAD system allows a graphics data base to be manipulated by authorizing the storage and retrieval of all the entities (points, lines, planes, curves, complex surfaces etc) which make up a graphics file. Basic graphics commands make up part of the CAD system software. They permit:

1. input of simple graphics entities (points, lines, planes, cones etc);
2. manipulation of these entities (enlargement, intersection, perspective definition, copying by line or point symmetry etc).

Commands for the movement of entities are useful in robotics (eg translation and rotation): they allow the configuration in an articulated mechanical structure to be designed. Another type of software available is known as the *interpreted programming language*, which enables the basic commands to be manipulated by programming. This operation can be carried out by a user with little experience.

To cope with the possibility of exchange between the many types of CAD system already on the market, an IGES standard procedure, which ensures efficient dialogue between data bases, has been developed but is used by very few systems.

5.2.4 DEFINITION AND VISUAL DISPLAY OF AN OBJECT

5.2.4.1 Definition of an object

To demonstrate fully the use of a CAD system in defining an object in the data base, it is useful to describe the process by which a plan of the object can be developed. Using a CAD system the object is designed as a three-dimensional model, intended as a virtual representation of the object in three dimensions. Unlike systems for computer-assisted drawing, the production of the image is not the sole application of this representation. There are a number of different types of three-dimensional model:

1. *line model* used when the coordinates of the vertices (X, Y and Z) and the lines joining them are to be stored in the data base (see Figure 5.3).
2. *surface model* which can store a complex surface using parameters and can be manipulated, for example, to allow calculation of lines

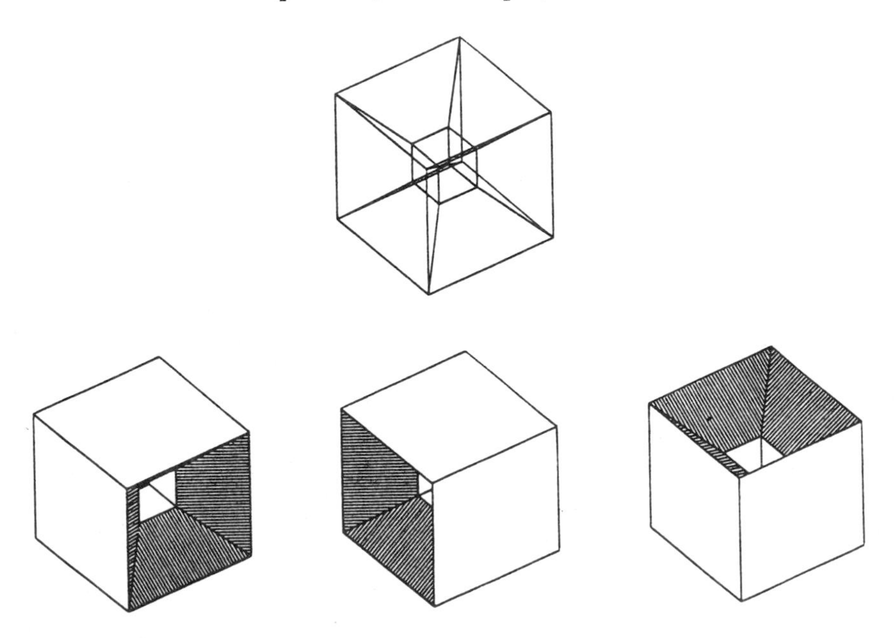

Figure 5.3. *Various possible interpretations of a line model*

normal to the surface or intersections with other surfaces. Various methods for surface approximation (eg Coons, Bezier, B-Spline) have been developed.

3. *solid model* which can be created if, in addition to the characteristics of a hybrid line and surface model, the concept of solidity is integrated into the model. Boolean operations used in relation to solids, such as calculation of intersection, union etc, should be applicable. A number of examples of solid models (eg Solidesign from Computervision, Euclid from Matravision, Catia from Dassault) allow a three-dimensional model to be developed from the object itself, either by breaking it down into elementary volumes or by construction based on primitives (see Figure 5.4).

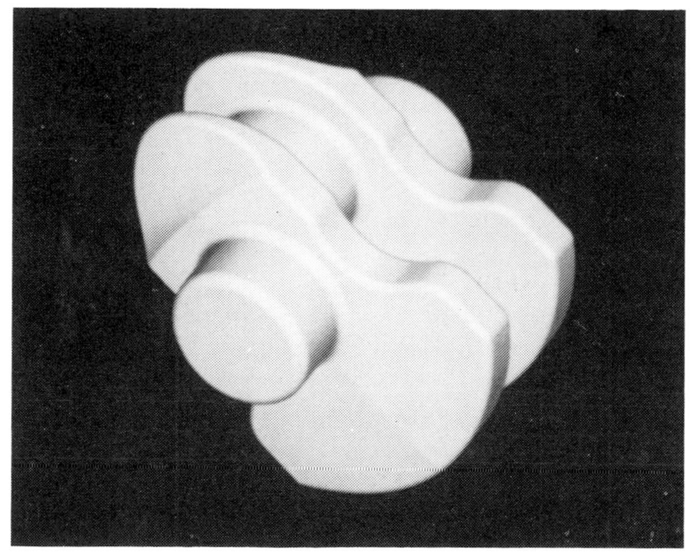

Figure 5.4. *Visualization of a solid model (from Computervision)*

The state of the art in CAD concerns line and surface models, which are already widely used, with good results, for the definition of objects and the programming of numerically controlled machines. The volumetric approach allows the use of a model with a large quantity of information to simulate or program many robotized tasks. Among these tasks, assembly is the one which merits the use of this type of model for its programming. Although the time required to get access to the data base and to carry out calculations in a CAD system is reduced, it is currently too long to allow generation of solid models that can be used to any good effect in programming and simulation. The approach used in practical applications is, therefore, limited at the moment to the line and/or surface method.

The input of data required for the formation of a model may be carried out explicitly: by giving the coordinates of the entities necessary for the development of the model, using an alphanumeric console. Using this method, a straight line would be defined by its two extreme points.

Another method consists of using graphics input to design the object by an interactive process, in the course of which the user will digitize or select the entities. The object itself may be broken down into a number of subsets, modified at different levels, thus allowing all or only part of the object to be displayed.

5.2.4.2 Screen display

Once the object has been defined, the user works on a graphics file, made up of a certain number of views of the object. The possibility of working simultaneously on a number of views can be envisaged. The view or views correspond to the standard view used in industrial plans: top view, front view, side view, cross-sections and perspective etc. This set of views is presented on the VDU in a conventional way (see Table 5.2).

top	side
front	perspective

Table 5.2.

The user defines the perspective he wants and a key pad can be used to make the object turn around as if being viewed by a camera.

5.2.5 METHOD OF USE

It has been shown that the use of graphics inputs (eg digitizing table) takes place interactively. By applying basic graphics commands or those developed by the user himself, it is possible to enter, in the form of program data, or during a command, graphics information provided by the user with the help of the digitizing table and so to develop a process in which the user can apply his intelligence and experience. The interactive approach (responding to questions, choices or indications given by the digitizing table) is at the root of the CAD system, whatever the process to be simulated or programmed. This is one of the advantages of using CAD and it should be emphasized in relation to robotic applications. This is discussed further later in the chapter. The presence of a *menu* inserted into a digitizing table assists the man-machine dialogue. This menu is made up of basic commands and those developed by the user as well as any simple (eg line, circle) or complex (eg cube, cylinder) geometric object. *Digitization* can be used with the contents of this menu to simplify the sequence of tasks necessary for the development of a geomctric model and to increase the speed and efficiency of programming or simulation work (see Figure 5.5).

Figure 5.5. *Touch tablet menu (from Computervision)*

5.3 Use of CAD methods

5.3.1 GENERAL POINTS

From the brief description given earlier of a CAD system, it can be seen that the following are available:

1. calculation power equivalent to that of a standard processing system (minicomputer with average power);
2. graphics functions for visualization and manipulation of geometric entities of a three-dimensional model using basic commands or commands developed by the user.

Three stages can be identified between the three-dimensional model of a robot and the generation of programs, which show in detail the advantages of using CAD in the development of an automated process:

1. modelling;
2. program generation;
3. simulation.

5.3.2 MODELLING

Whether it involves simulating robot movements or obtaining graphics information for use in programming proper, it is vital to obtain a geometric model that can be easily used by the robot, at an early stage. This definition will relate to the objects involved in the task to be carried out by the robot as well as the environment in which it will take place. At this stage of modelling, the preliminary phase, is a static definition of the elements involved: robot, objects and environment.

5.3.2.1 Static definition

5.3.2.1.1 Robot

At the elementary design level, a robot is represented as an articulated system, that is a number of lines linked by translational or rotational axes. By this method, a line drawing representation of the robot is obtained. This contains the essential elements of definition: distances between axes, distances along or about which movements are carried out. The AKR 3000 robot, for example, would be represented as shown in Figure 5.6, respecting the AFNOR norm.

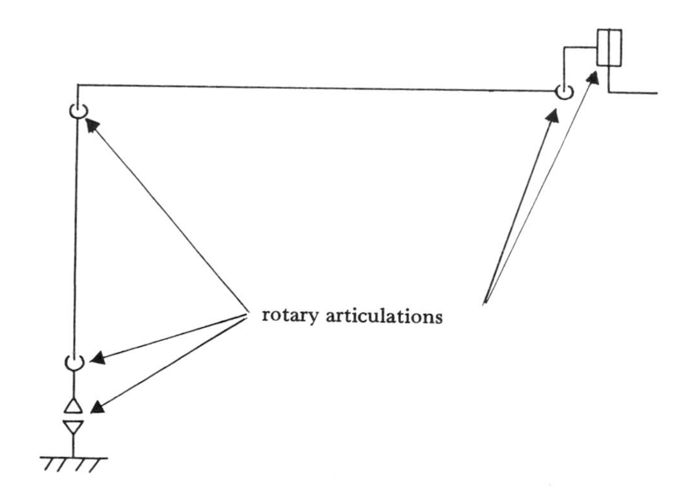

Figure 5.6. *Normalized representation of the AKR 3000 robot*

It is possible to add significant information at the programming level at which the robot design can be defined interactively:

1. dimension tolerances for the lengths of the lines and interaxial parts;
2. linear and angular ranges of axis movement.

The software developed, for example, at LAM in Montpellier on the Catia system, and CAE on the Calma system and at PSA on Computervision, helps the user to model and validate the characteristics of a robot. The advantage of such a method of modelling lies in the parametrization of the variables defining the robot. It can be used to update the robot model at any point.

A more precise method of geometric modelling consists of attaching volumes to the skeleton defined in the model, so as to obtain an idea of the real bulk of the robot. By using simple geometric volumes (truncated cones, cylinders, prisms etc) it is possible to develop a robot model which takes account of the work envelope (see Figure 5.7).

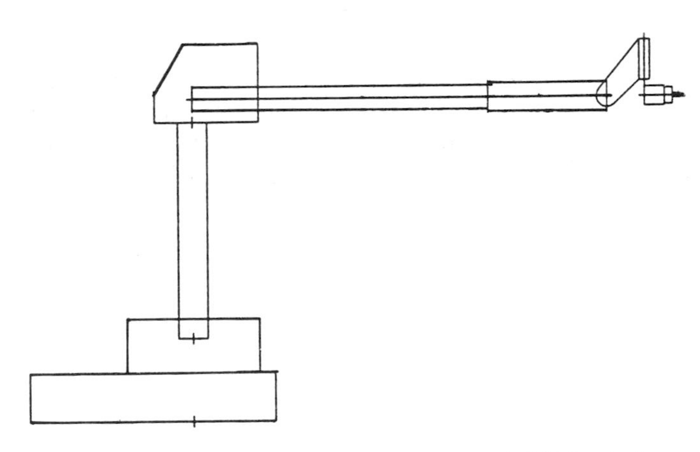

Figure 5.7. *Geometric model of the AKR 3000 robot*

As mentioned in Section 5.2.4.1, this type of model is either a line model or a hybrid line-surface model. Using software for robot definition can provide the following information, in addition to all the applications available:

1. definition of the work space;
2. comparision of these volumes for different robots (eg by superimposing them on the VDU);
3. designation of the best adapted end effector;
4. comparison between production methods for two robots of the same basic type.

5.3.2.1.2 Object

Defining an object in a CAD system is extremely simple. Compared with the problems that can arise when modelling is attempted on standard computer systems using syntactic methods (textual and analytical description), the facility offered by CAD systems for the same purpose can present a tremendous advantage.

If the objects to be manipulated or otherwise involved in the robot task have already been designed using a CAD system, the problem of modelling no longer arises. This is the case when the objects are the result of a machining phase designed on the CAD system. Thereafter, these objects, although modelled already, must be prepared to allow the desired program to be formed: a car part designed using a CAD system must also contain the position of welding spots and an indication of lines perpendicular to these points before it can be assembled using spot welding. The most important problem at this stage of model formation is the relative positions of the sets of coordinate axes for the definition both of the robot and the object. This is discussed in detail in Section 5.3.2.2.

5.3.2.1.3 Work space

The method used to define the work space is similar to the method described to define the robot and objects geometrically. The environment or work space is either fixed or flexible. In the first case, when the application relates to an installation, a feasability study of the environment is considered or a study of collision problems made. The geometric forms can be simplified to a menu figure, for example, that the user can call as required. The second case which relates particularly to robot production lines is based on defining the object with the emphasis on the precision of geometric shapes and the installation.

5.3.2.2 Formation of the model

Section 5.3.2.1 explained the method by which the geometric elements (robot, objects and environment) which feature in the model of the task are separately defined. Defining the model takes place through the installation of the various reference points related to the elements. The contribution of CAD is therefore important. In a preliminary feasability study the CAD system allows the objects, or as is more frequently the case, the robot in relation to the objects to be positioned as a function of the accessibility of these objects. The visual display of the work space allows this relative position to be defined such that the objects are situated in the work space. This first approximate definition of the positioning may, in some cases, be refined if the robot articulated variables are known.

5.3.3 CALCULATION OF ARTICULATED VARIABLES

The problems associated with the calculation of these variables are described in Chapter 3 (Section 3.5.3), and correspond to the solution of an equation system given by the geometric model relationship:

$$X = F(q) \qquad \textbf{(5-1)}$$

where X is a vector with n components expressed in a set of Cartesian coordinate axes (n = number of articulated variables), and q is a vector with n components (articulated variables). The method by which this equation is obtained and solved are described in Volume 1 of this series entitled *Modelling and Control* (Coiffet, 1983). Calculating the articulated variables on a CAD system relies on:

1. defining X;
2. applying the method of solution.

5.3.3.1 Definition of a configuration

Object models, as already explained in Section 5.3.2.1, are prepared for the task to be automated. More precisely, the graphics file containing the object model also contains the geometric information necessary for the definition of the various robot configurations during the development of the task.

The vector in equation (5-1) is therefore an implicit element of data in the CAD system. In a point-to-point welding task, for example, employing a robot with six DOF the user digitizes the position of the spot to be welded as well as the direction of welding. In such examples requiring a high degree of precision, the condition of perpendicularity must be respected. The software supplied with the CAD system allows straightforward calculation of the lines perpendicular to a surface defined in the system, with the help of various methods already mentioned in Section 5.2.4.1. The processing system deduces the six components of vector X precisely. The advantage of using CAD rather than a training method is obvious.

The work carried out at PSA on the Barnabe robot, or at Automatix on the Computervision system, uses this definition of robot configurations for the kind of welding operation described above. The same applies to a washing task carried out by an AKR 3000 robot in a FMS at the Citroen factory in Meudon (discussed in detail in Section 5.4.2.2).

In all cases, a program for the calculation of articulated variables allows the components of vector X to be input, and these can be used either in the data base, or interactively, by identification on the VDU of the elements stored (ie points and directions).

5.3.3.2 Geometric solution method

When using the required configuration, the CAD system must be able to solve the general relationship X = F(q) in order to calculate the articulated variables. Many methods can be used: calculation by structural analysis consists of determining the rotational or translational centres by intersection of the spheres or planes and spheres, starting from a known position of the end effector, until all the positions of the robot sets of coordinate axes are known, and thus the articulated variables are found. The values of the variables are given in relation to reference points or initialization configuration which, in the model, corresponds to a zero position of the robot encoders. It is, therefore, possible that at this level there will be differences between the model and the real system. The drawback of this method is that it can only be applied to a particular type of robot. It can, however, sometimes be reused with other robots, because of the degree of similarity between some robots and the possibility for parametrization at model level (see Section 5.3.2.1).

Whatever the method applied, the solution of problems relating to the calculation of articulated variables will result in either no solutions (*inaccessible configurations*) or in one or more solutions which are not allowed because of software limits (*inaccessible configuration due to range of movement*) or allowed, if the n values of the components of q satisfy the constraints of the range of movement (*accessible configuration*).

5.3.4 INSTALLATION

Hence, it can be seen that it is possible to define the position of the robot in relation to the objects involved. There may be many criteria. The first to be considered is that of the accessibility of all the required robot configurations. It is also possible to attempt modification of this new relative position, so as to avoid excessive torsion in the wrist which could rupture supply cables or those connecting fixed sensors to the end effector. The installation of robots in the PSA group has been carried out using this method since 1981.

In situations in which objects move relative to the robot (or vice versa) during the task, it is possible to take these movements into account as long as the analytical laws governing these movements as well as those of the robot (velocity, acceleration etc) are known. The most common situation involves robots working on a production line (eg assembly, spot welding, paint spraying).

These comments are also valid for geometric modelling of the task in order to carry out programming or simulation.

5.3.5 CONCLUSIONS

The methods used for forming a geometric model of a task based on the static definition of the elements have already been discussed. The advantages of using CAD are that the geometric elements can be defined with precision, and the real installation of these elements, while satisfying the constraints of configuration accessibility, in particular, can be defined. Without the help of CAD the work can be lengthy and difficult for the designer. Differences between the model and the real system can arise in relation to:

1. the geometric definition (inaccurate dimensions of segments, approximation of the volume of objects, using work envelope surfaces etc);
2. installation of the robot, if this is carried out in the model by copying plans and above all if the installation must be modified on site.

In this situation, it is possible to perform a trial test of the task to allow the relative positions of the elements with respect to the robot to be stored in the memory. This on-line modification is not always possible in industry. It can be obtained by unlocking the zeros on the axes or by modifying the work reference frame in some robots. In other models, all the points generated by CAD must be corrected on site.

5.4 Program generation

5.4.1 INTRODUCTION

The off-line use of models for programming purposes is not widespread. The methods which do exist at PSA and Automatix, for example, demonstrate the advantages derived from the use of CAD in robot programming. Classifying these methods, and the various research programs which are described later, into levels of complexity of 'intelligence' presents certain problems. An attempt is made to define the aim of the generation of point trajectories (see Section 5.4.2) and the generation of textual programs (see Section 5.4.3).

5.4.2 GENERATION OF POINT TRAJECTORIES

The aim of this type of programming is to generate using a CAD system a code that can be assimilated by a playback robot:

1. movement of the robot between successive configurations during the task according to the code;
2. execution of commands involved in the task (eg opening gripper, timing, start and end of welding operations).

The code for assimilation by the robot may be machine code used in the standard training robot described in Chapter 3. In this situation, the procedure for inserting such programs into the memory must be perfectly mastered. The other solution is to generate an intermediate code such as IRDATA (see Chapter 4), if this can be understood by the robot.

By analysing the nature of the task and the characteristics of the robots, the classification used in Chapter 3 can be applied to describe successively:

1. point-to-point trajectories;
2. calculated trajectories (ie point-to-point with interpolation of the trajectory);
3. complex trajectories.

5.4.2.1 Point-to-point trajectories

The possibility of calculating robot articulated variables using the geometric data present in the graphics file of the object model has already been considered (see Section 5.3.3). For a set of points corresponding to accessible configurations it is, therefore, possible using the CAD system to generate a file containing either the angles between the different axes, which may be direct encoder values or incremental, for each of the configurations, or else the n components of the vector of the configuration in the set of Cartesian coordinate axes (ie three coordinates of position and three Euler angles), if the control system includes a coordinate transformer.

Apart from positional information, the file contains the conditions for movement from one point to another:

1. timing;
2. waiting for an external event, sending signals to the environment.

The facilities provided by the editing of such a program on a CAD system are the same as those obtained by storing the program in the central memory (see Section 3.4.1). CAD offers an extra advantage in that it retains the visual link between point n in the program editing file and its visual position on the object (as in spot welding) in the memory, which allows modification of insertion of points (see Figure 5.8).

CAD also allows two other important aspects of on-line programming of point-to-point control robots:

1. trajectory design, allowing collisions between trajectories and objects to be taken into account;
2. performance calculation.

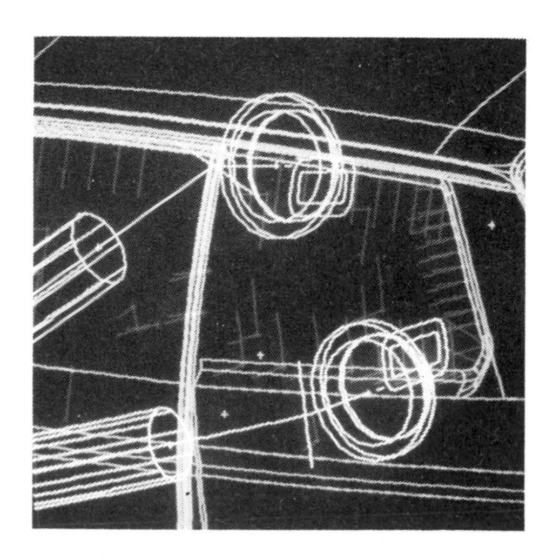

Figure 5.8. *Spot welding by the Barnabe robot (from Peugeot-Citroen)*

If the control laws of the actuators are known, it is possible to study these two problems on a CAD system. For a point-to-point control robot in which the trajectories are not calculated, however, designing the trajectory between two distant points is difficult: the trajectory is not mastered at the control system level. In the same way, time calculation is not easy: the control laws depend on the servo-systems and actuators and are not constant. Although these problems are discussed in this volume, a detailed description is given in connection with point-to-point control robots with calculated trajectories only.

5.4.2.2. Calculated trajectories

In CAD systems, the possibilities presented by the use of this type of robot are the same as those described in Section 5.4.2.1 with the addition of data on the type of trajectory followed and the velocity of tool movement achieved between the points. Apart from allowing the generation of trajectories in code form, this information enables the design of trajectories and the calculation of performance times to be determined.

The example of the application developed for the AKR 3000 robot in the FMS at Meudon illustrates the various phases in the programming process of a CAD system (supplied by Computervision). This robot with five degrees of freedom (skeleton in Section 5.3.2.1) has a terminal device made up of nozzles. These nozzles are cylindrical tools of different diameters used to squirt either water or air into machined holes, in cases where traditional washing or drying methods are unable to remove shavings and lubricating fluids from the machining process.

The work points (ie those at which the fluid is squirted) are featured in the object model in terms of:

1. the position (X, Y, Z) of the end effector;
2. the direction (Euler angles: H, L) of the end effector.

The work points are stopping points for the robot. With each one of the work points (T) an approach point (A), in the direction of the tool away from the part, is associated at a certain parametric distance (d). The robot must pass through these points with a specified speed. The trajectory (A, T) is performed slowly. In addition to these two types of point A and T, it is possible to insert avoidance points (E): the robot passes nearby but the speed at these points is unspecified.

The object, a machined mechanical part whose volume is contained within a 50-cm cube is fixtured on a pallet, the pallet is translated to a rotary table which can take eight indexed positions about a vertical axis. With the help of this possibility for object rotation and the calculation of the articulated variables at each of these points, it is possible to create a file for input made up of work and approach points, which correspond to the accessible configurations. Since the robot under consideration carries out a linear trajectory between two points (work or approach), the software developed at PSA on Computervision equipment allows these trajectory lines to be traced, and modified to avoid collision, in the following way:

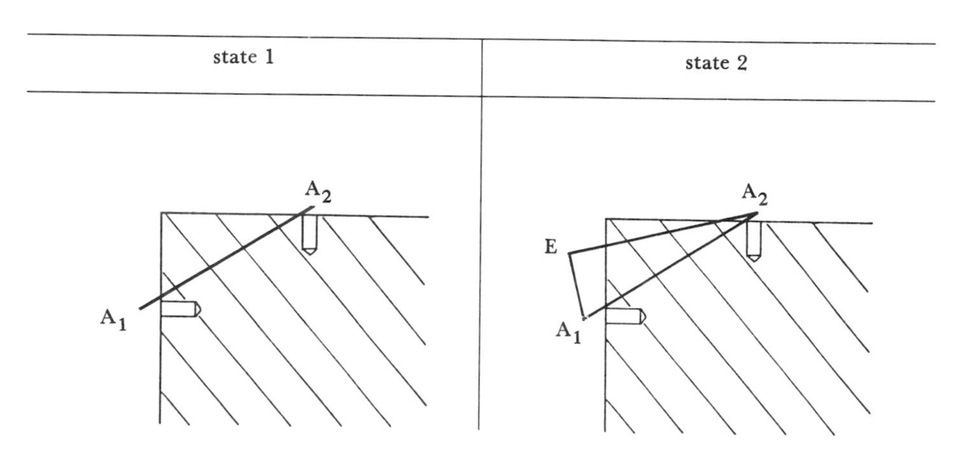

State 1

– Reprocessing of two consecutive approach points in the input file.
– Line of trajectory (A_1, A_2). *Question*: 'Do you want to modify the trajectory?' *Answer*: 'Yes'.

State 2

– Stylus pointed at screen (point E).
– New trajectory (A_1, E, A_2) if E is accessible.
– *Question*: 'Do you want to modify the trajectory?' *Answer*: 'Yes'.
– *Question*: 'Do you want to retain the added point?' *Answer*: 'No'.

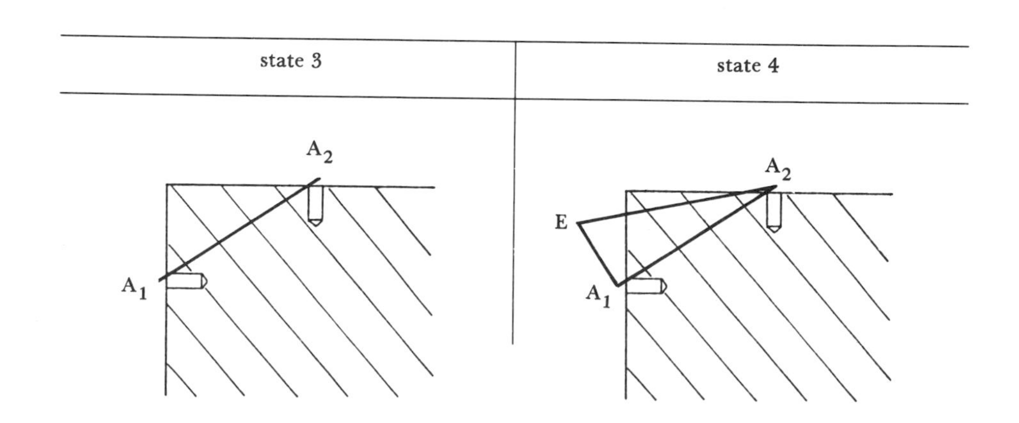

State 3
– Return to state 1.

State 4
– New point indicated, then traced.
– *Question*: 'Do you want to modify the trajectory?' *Answer*: 'Yes'.
– *Question*: 'Do you want to retain the added point?' *Answer*: 'Yes'.

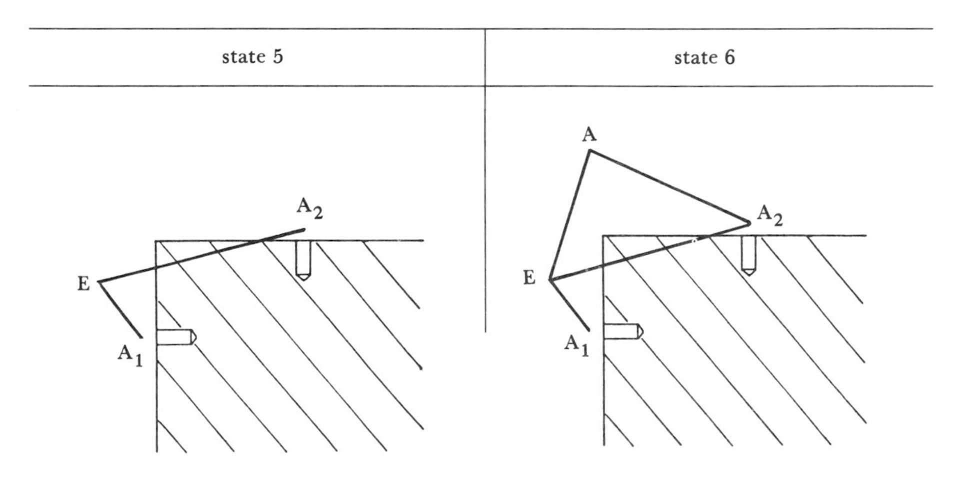

State 5
– Point E written in output file following A_1.
– Deletion of old trajectory (A_1, A_2).

State 6
– Stylus pointed at screen (point A).
– New trajectory (A_1, E, A, A_2) if A is accessible.
– *Question*: 'Do you want to modify the trajectory?' *Answer*: 'No'.

At state 7 (the final state) point A is written in the file, the different lines are deleted and there is movement to two approach points following (A_2, A_3):

Starting with a file (F_1):

. . . .
A ⊔ X . . . Y . . . Z . . . H . . . L . . . (point A_1)
A ⊔ X . . . Y . . . Z . . . H . . . L . . . (point A_2)

A new file (F_2) is developed:

A ⊔ X . . . Y . . . Z . . . H . . . L . . . (point A_1)
E ⊔ X . . . Y . . . Z . . . H . . . L . . .
A ⊔ X . . .
A ⊔ X . . . (point A_2)
. . . .

The direction of the tool for the interpolated points is the same as that of point A_1. An indication on the screen supplies two coordinates, the third (depth) is taken from A_1.

Using the method described above, the collision test is visual and the interactive procedures of a CAD system are used for modification of the trajectories as they are developed. The problems encountered are mainly:

1. the difficulty for the operator of integrating all the views available on the screen to estimate the risk of collision;
2. the necessity on a system, such as the Computervision Cadds 3, of progressing to the insertion of points in a view which is said to be active, but which is not necessarily the most convenient for the operator.

Despite these problems, it will be seen (see Section 5.5) that the design of the robot trajectories in each configuration and possibly between these configurations, in conjunction with the method of insertion of collision points and/or approach points, will allow an output file to be obtained, containing only:

1. accessible configurations linked by;
2. trajectories which avoid collision.

This output file also contains a number of commands relating to the execution of the task.

Figure 5.9 shows a code generated by the Computervision system for programming the AKR 3000 robot. The commands featured in this example are:

⟨R⟩⟨n⟩: table rotation, n from 1 to 7;
⟨V⟩: open nozzle;
⟨K⟩⟨m⟩: timing of n tenths of a second.

5.4.2.3 Complex trajectories

It has been shown that using CAD for programming robot applications allows:

1. the required robot configurations to be achieved (see Section 5.4.2.1);
2. the movement from one configuration to another to be validated or modified by trajectory design and visual detection of possible collisions (see Section 5.4.2.2).

The technology currently used in robotics allows the information necessary for a point-to-point control program for tasks such as loading and unloading, spot welding and washing to be stored in the central memory (using buffer memories if needed).

```
10!K   50
11!A   X748       Y-2848    Z2480      H9000    L0
12!E   X+02456    Y-02848   Z+02655
13!A   X1849      Y-249     Z2980      H18000   L0
14!T   X849       Y-249     Z2980      H18000   L0
15!K   50
16!A   X1849      Y-249     Z2980      H18000   L0
17!F
18!E   X+01331    Y-00248   Z+04515
19!A   X-7200     Y-13100   Z4500      H0       L-9000
20!R   2
21!V
22!A   X457       Y1249     Z4263      H0       L-4500
23!T   X1165      Y1249     Z3556      H0       L-4500
24!K   50
25!A   X457       Y1249     Z4263      H0       L-4500
26!A   X-750      Y-770     Z4600      H9000    L-9000
27!T   X-750      Y-770     Z3600      H9000    L-9000
28!K   50
29!A   X-750      Y-770     Z4600      H9000    L-9000
30!F
```

Figure 5.9. *Example of the intermediate code generated on Computervision equipment for a washing robot in the FMS at the Citroen factory in Meudon (from AKR Peugeot-Citroen)*

Many tasks do not fulfil the point-to-point criteria. The best known are paint spraying and continuous welding of complex parts. In painting operations, the robot is frequently required to copy complex movements with no simple kinematic law to follow, which are, as a result, difficult to program.

A weld joint corresponds to a curve that must be defined or approximated analytically using a CAD system, by intersection of two surfaces. Determining the geometry of the weld joint is completed by a knowledge of welding process. This process takes into account various parameters: velocity of the end effector, the thickness, nature and treatment of the metal plates etc. The values of these parameters, which are fixed during a period of interactive use of the CAD system, allow the following to be determined:

1. number of points sampled in the curve;
2. position of the end effector at each of these points.

After the accessibility of these configurations has been verified, the CAD system can generate a program written in intermediate code which will be stored in the robot memory. The contribution of CAD to this example of continuous welding arises from the analytical definition of the trajectory (T) and constraints on this trajectory (Q) (see Figure 5.10).

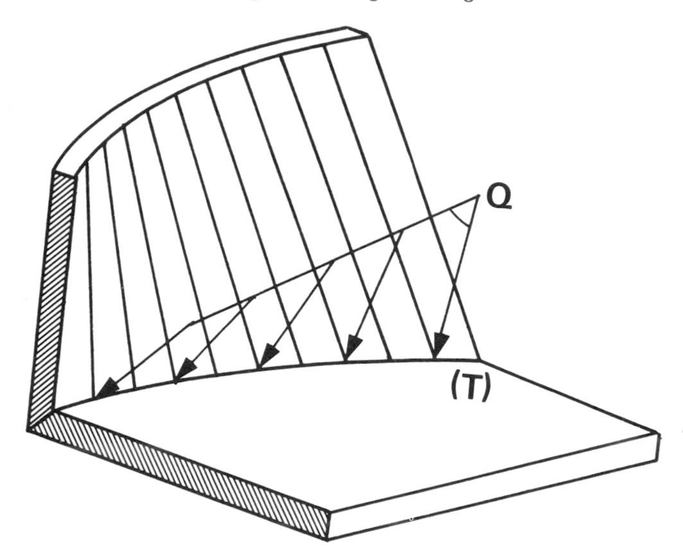

Figure 5.10. *Programming seam tracking*

Since the number of points to be programmed is large, a method for programming with CAD offers off-line not only the advantage of geometrically defining the configurations accurately, but also saves a significant amount of time.

5.4.3 GENERATION OF TEXTUAL PROGRAMS

Robot programming at this level is based on the description of the task in algorithmic form and in terms of the robot movements specified generally in terms of reference frames. This description, which is independent of the mechanical structure of the robot, involves the implantation of a language into a minicomputer-type processing system. The system must offer a number of fundamental possibilities:

1. generation of robot movements through a coordinate transformer;
2. interaction with a data base consisting of various frames;
3. execution of a program in a non-linear form.

The main contribution of CAD to programming of this type is that it allows a language to accept information from a preliminary modelling operation for the task using a CAD system. In particular, the painful computation of frames can be greatly simplified, collisions can be forecast and simulations can be performed to optimize the program. However, the generation of textual programs based on a CAD system is still at the research and development stage.

5.4.3.1 Automatic program generation

This concerns a line of research which was carried out on the RAPT system at the University of Edinburgh and on the LM-GEO at Imag in Grenoble. At the University of Edinburgh a CAD system has actually been used (Computervision Cadds 3), but at Imag a new graphics processor is being developed.

Generally speaking, the automatic generation of programs is intended to provide an end effector level program based on a description of the task at the object level. The end effector program may, for example, be written in VAL or LM for a task description using RAPT or LM-GEO respectively.

VAL requires the on-line specification of the various robot configurations in the course of the task. The RAPT system, however, allows a task to be programmed using VAL in conjunction with a CAD system for the definition of configurations (Ambler *et al.*, 1982). Figure 5.11 shows programming with the RAPT/Cadds 3 system, in which the positional information is derived from a model of the task. Apart from the characteristics specific to the language in use, this example can be assimilated and used by the example of CAD given in Section 5.4.2.2, with the code generated on a Computervision system for an AKR 3000 robot. For the same example, assembly programming, LM-GEO (Mazer, 1982), would require a syntactical description of the objects and their relative positions, which is difficult to apply. However, by applying artificial intelligence techniques, progress towards automatic program generation using high-level languages is promising for the future.

5.4.3.2 Computer-aided program generation

As seen in the previous example, a programming language can utilize the potential of a CAD system. The menu which appears on the digitizing table may, in this connection, prove to be a powerful tool for mixing the instructions in a given language with the geometric information stored in the object models, in the process of generating robot motions (see Figure 5.11).

This generation of movement is associated with the instruction of the MOVE type:

– MOVE P1	VAL
– MOVE (4,5, 30,90)	AML
– MOVE OBJECT (3) TO OBJECT (4)* SITUATION (4)	LM
....	

Modelling a task on a CAD system has the advantage of allowing instructions associated with the movements to be validated by off-line determination of the configurations, which, among the required

BENCHMARK LEVEL 2 – OFFLINE VAL WITH RAPT/CADDS 3 DATA

```
.PROGRAM LEVEL 2
   1. OPENI 0.00
   2. REMARK FETCH THE BLOCK AND SLIDE IT INTO THE ASSEMBLY JIG
   3. GOSUB BLOCK
   4. REMARK FETCH THE PLATE AND SLIDE IT ONTO THE BLOCK
   5. GOSUB PLATE
   6. REMARK FETCH THE PIN AND INSERT IT INTO THE HOLES
   7. GOSUB PIN
   8. STOP
.END
```

```
.PROGRAM BLOCK
    1. MOVE P21
    2. SPEED 20.00
    3. MOVES P20
    4. DELAY 1.00
    5. CLOSE I 0.00
    6. DELAY 1.00
    7. MOVE P21
    8. MOVE P22
    9. SPEED 20.00
   10. MOVE P23
   11. SPEED 10.00
   12. MOVES P24
   13. DELAY 1.00
   14. OPEN I 0.00
   15. DELAY 1.00
   16. MOVE P22
   17. RETURN 0
.END
```

```
.PROGRAM PLATE
    1. MOVE P31
    2. SPEED 20.00
    3. MOVES P30
    4. DELAY 1.00
    5. CLOSE I 0.0
    6. DELAY 1.00
    7. MOVE P31
    8. MOVE P32
    9. SPEED 20.00
   10. MOVE P33
   11. SPEED 10.00
   12. MOVES P34
   13. DELAY 1.00
   14. OPEN I 0.00
   15. DELAY 1.00
   16. MOVE P32
   17. RETURN 0
.END
```

```
.PROGRAM PIN
    1. MOVE P11
    2. SPEED 20.00
    3. MOVES P10
    4. DELAY 1.00
    5. CLOSE I 0.0
    6. DELAY 1.00
    7. MOVE P11
    8. MOVE P12
    9. SPEED 20.00
   10. MOVE P13
   11. SPEED 5.00
   12. MOVES P14
   13. DELAY 1.00
   14. OPEN I 0.0
   15. DELAY 1.00
   16. MOVE P12
   17. RETURN 0
.END
```

POSITION	X	Y	Z	O	A	T
P10	−379.4	444.6	−386.3	124.8	11.1	−44.5
P11	−379.4	444.6	−275.2	124.8	11.2	−44.4
P12	15.5	−656.1	−58.0	126.1	41.4	−36.9
P13	206.4	−609.4	−229.7	126.1	41.4	−36.9
P14	234.9	−602.4	−255.4	126.1	41.4	−36.9
P20	25.9	454.2	−253.9	165.0	90.0	0.0
P21	25.9	454.2	−142.9	165.0	90.0	0.0
P22	345.2	−575.3	46.8	57.4	48.8	33.4
P23	388.6	−564.6	−202.0	57.4	48.8	33.4
P24	345.2	−575.3	−253.2	57.4	48.8	33.4
P30	−168.2	517.8	−305.9	−178.0	90.0	0.0
P31	−168.2	517.8	−194.9	−178.0	90.0	0.0
P32	344.4	−575.5	85.4	57.4	48.8	−33.4
P33	344.4	−575.5	−214.6	57.4	48.8	−33.4
P34	316.6	−582.4	−247.3	57.5	48.8	−33.4

Figure 5.11. *Generation of VAL program by the RAPT/Cadds 3 system*

positions, are effectively accessible. The other advantages of CAD (rapid construction of the data base containing positions and orientations, precision, ease of modification, collision tests etc) have been discussed in Section 5.4.2. Another example of CAD application (Automatix) in which these advantages appear in the framework of programming in RAIL language follows. CAD has allowed off-line definition of the points of insertion for electronic components in a printed circuit board, the preliminary design of which has been carried out on the CAD system. The number of points to be defined and the precision with which they must be located makes any method other than CAD (training, sensor vision, direct input of coordinates etc) lengthy, imprecise or tedious.

Another example developed by Automatix allows modification of a program designed by CAD, by taking into account external information. This method is used for continuous welding (see Section 5.4.2.3). The master control system allows input of a difference in calculated positions into its trajectory calculation module:

$$(\underline{P}) = \underline{Pe} - \underline{Pc} \tag{5-2}$$

Pc → position delivered by visual sensor
Pe → expected position (CAD)

In this way, a feedback robot operation is obtained.

5.4.3.3 Note on task modelling

Complex tasks using language often require more elaborate task modelling than simply generating the trajectory with points, for example:

1. trials in which the object model must include projected surface, centre of gravity, perimeter of exterior contour characteristics etc.;
2. assembly when the precise position of two objects in relation to each other is required.

At the moment geometric modelling on CAD systems is not fully satisfactory for these tasks. In the case of assembly, the determination of the necessary movements (rotation and translation) for positioning the objects is not made any easier by the use of a CAD system because there is no model describing the surface and the volume (Ambler *et al.*, 1982). A syntactic description, of the type used by LM-GEO, seems more appropriate, at least at the present, to the solution of this type of problem which is so characteristic of assembly (Mazer, 1982, 1983). More advanced CAD systems than those currently available on the market are required to help solve this type of problem. These systems should allow the movement of solid objects to be displayed.

5.5 Simulation

5.5.1 AIMS

Simulation based on robot behaviour is made up of an *animated cartoon* of the robot configurations during the course of the task. Notions relating to time (length of time required by the robot to move from one point to another, to complete a trajectory or stopping time at a given point etc) or simulation of the behaviour of sensors etc can be incorporated into the basic simulation.

5.5.2 IMPORTANCE

Simulation is important in that it can help solve optimization, implantation and collision problems. These aspects are discussed in detail in Sections 5.3.4 and 5.4.2.2. Simulation can also help solve problems of a theoretical type, such as determining optical trajectories. It can also reduce costs, pilot studies and installation time. In a more general sense whatever the applications under consideration may be (whether there is risk of collision or not), all robot users agree that simulation of robot tasks is an invaluable aid to programming. Given the variety of types of robot (simple handling devices, painting robots with seven axes, Cartesian robots etc) the user of robots will appreciate the help of a simulation tool which will allow him to avoid working blind on a robot he knows very little about. These ergonomic considerations are all the more important in that current research into programming languages (see Bonner and Shin, 1982) is tending towards the development of a language that can be understood and handled by users with a variety of experience and competence.

5.5.3 METHODS

Consider a point-to-point control robot in which the trajectory is not mastered (see Section 5.4.2.1). It is assumed that the actuator control laws are known. In such a robot, the configuration at time t of the task is known, and therefore can be entered into the design.

For all robots, the trajectory can be found in this way, as can the kinematic law of the trajectory. It should be noted that at the avoidance points (see Section 5.4.2.2) the configurations can only be estimated. The methods used for establishing simulation are simple:

1. all the programmed configurations of the robot are designed, one after the other, using the commands for translation and rotation in the CAD system;
2. the above process is carried out using a program written in interpreter language (Varpo for Computervision) or, for example, with commands written in FORTRAN.

A program for trajectory interpolation may, if necessary, accommodate visual display of the intermediate robot configurations. For a visual collision test, the problems encountered are of the same order as those mentioned in Section 5.4.2.2: the absence of algorithms, elimination of hidden lines, and problems of integration, by the user, of the different screen views make these visual collision tests difficult (see Figure 5.12).

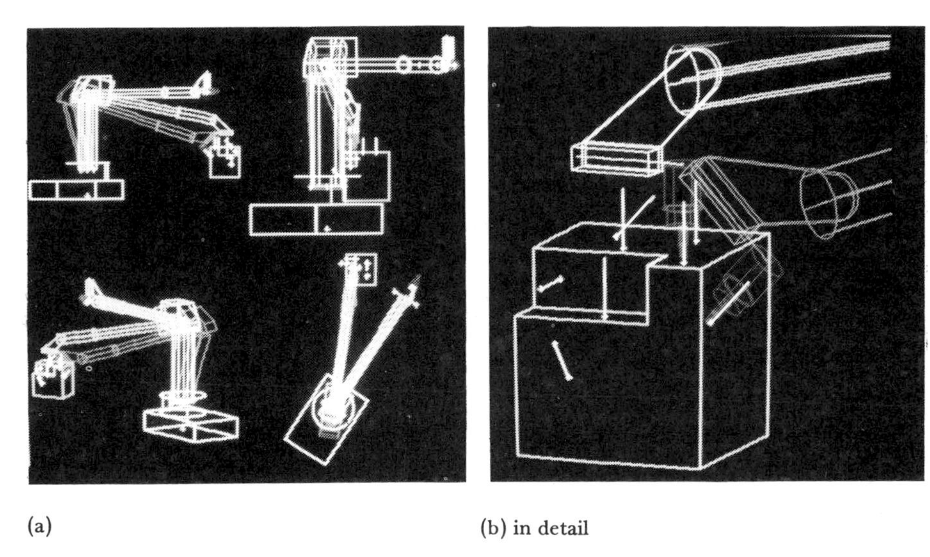

(a) (b) in detail

Figure 5.12. *Simulation of robot movement*

Real-time simulation, required by some users to allow improved understanding of dynamic and real-time problems in robots, is still not available in standard systems. The calculation time is too long, even with the most efficient systems currently available, even with line models to allow images to be generated in less than 1/30th of a second. If, moreover, the user requires a real image (solid modelling), the calculation time can be of the order of a minute per picture, which rules out the possibility of any dynamic simulation. Despite this, the progress made in systems of image synthesis in other fields (cinema, flight simulators etc) is encouraging for the possibility of real-time simulation in the near future.

5.6 Industrial application

5.6.1 IMPLEMENTATION OF PROGRAMS DEVELOPED OFF-LINE

It has been shown that programming an automated task on a CAD system theoretically allows the training phase to be avoided. However, the user may be required during the installation phase. The task may be programmed while the robot is in use and program (n + 1) may be loaded during the execution of program (n). This teleloading is carried out either directly from the CAD system to the robot (as in the Barnabe robot used in PSA factories) or indirectly via another processing system (eg the solution used in the Citroen FMS in Meudon, in which the link between the Computervision system and the AKR 3000 robot is the central control computer Solar 16/65).

Of course, teleloading can be dispensed with when the robot task is repetitive; the saving in time occasioned by programming off-line is still significant as compared with programming by training, which can hold up the progress of the operation in which the robot is integrated for several days.

5.6.2 CAD, NUMERICALLY CONTROLLED MACHINE TOOLS AND ROBOTS

Considerable efforts are currently being made by the manufacturers of CAD turnkey systems to develop software specifically for robotics. In the same way, the designers of numerically controlled systems are beginning to turn towards robotics (eg NUM and General Electric). A number of developments such as Robex (Taylor *et al.*, 1982; ISIR, 1981) or MCL (ISIR, 1981) can be situated in the context of a tentative integration of robots in industry which already uses numerically controlled (NC) modules with languages compatible with ATP. Generating programs for NC, using CAD stations, has been used in many industrial applications since the early 1970s. The description of the task, made using an interactive process, allows the definition of the paths to be followed by the tool, the forward speed characteristics and spindle rotation etc.

In the case of APT (Automatic Programmed Tool) language supplied with a CAD system, the CLDATA file in standardized code, which is generated in the system, is processed by the post-processor (or macro coder), which forms the interface with the numerically controlled unit (see Figure 5.13).

object model
CAD
APT program
CLDATA
post-processor
punched tape

Figure 5.13. *Generating a program in APT*

APT is a language which has not been defined for interactive uses. It has the advantage of relating, in principle, to all problems of numerical control, whatever the type of work envisaged (eg machining, cutting) and whatever the machine considered (three, four or five axes etc).

The absence of constraining specifications at language level, for establishing command and control of the processes and machines, is also a major drawback, since it results in interpretations of the language which vary depending on the process and machine under consideration. As a result, the automatic generation of numerical control languages must either be adapted according to the application considered or be reduced to a set of semi-automatic procedures. The current development towards the automation of processes involving numerical control is marked by the need to specify these processes so as to:

1. reduce and simplify interactive dialogue with the user. A physical model of the object should allow the user to define his choice, or at least to orient it somewhat; the repetitive sequences of the process must be easy to parametrize etc;
2. integrate the manipulation of the tools;
3. ensure effective dialogue and interfaces with the process as a whole (eg loading and unloading of parts, tools).

The experience gained in the field of NC and the developments in progress must be taken into account in assessing the possible development of program generation using a CAD system. The choice of language (or languages), of the method used for implantation and of a possible intermediate code etc, will condition the development of robotics associated with CAD. As with NC, the introduction of CAD into the field of robotics raises technical problems as much as problems relating to the organization of work: knowledge of the robotic process (see Section 5.4.2.3), the degree of accessibility and legibility of various levels of the program for potential users with varying levels of expertise, are just a few of the problems to be solved before a CAD system and robotic process can be fully integrated into industry.

5.7 Conclusions

CAD systems, although not designed specifically for robot users, can offer solutions to off-line programming problems, with a number of advantages:

1. a geometric data base defining the robot configurations can be rapidly and accurately generated and easily modified;
2. attainable configurations can be determined, and thus the program and implantation can be modified or corrected;
3. collision problems can be solved more readily.

The solution of this last problem is aided by the present state of CAD systems, which allow rapid generation of line drawings for simulation. If future CAD systems can:

1. generate colour images;
2. provide intelligent terminals;
3. generate and modify solid models;

then realistic simulations should be produced in real time without delay. The contribution of modelling and simulations is a significant aid in the development and modification of programs developed off-line.

The interactive operation of a CAD system should assist program writing in high-level languages. This generation may take, as its principle, the same attitude towards integration as that adopted for NC, which could result in the same acceptance in the sphere of automated processes.

Whatever the required level of programming, it is important to stress the potential offered by a robot processing system to modify on-line the errors of off-line programming.

These errrors arise from:

1. off-line programming errors (bugs);
2. differences between the model used off-line and the mechanism used in the factory (eg non-coplanar objects, imprecise positioning of the objects relative to the robot);
3. non-repeatability of the same program on two identical robots.

The solutions can be:

1. off-line: requiring complete knowledge of the robot characteristics;
2. on-line: a succinct training phase, a trial run, use of sensors etc.

References

ACMA *The ACMA Manual.*
AFRI *Definition and Statistics*:
AFNOR E 61-100 *Industrial Robot – Definitions for Mechanics, Geometry, Control and Programming* August, 1983.
AFNOR E 61-101 *Industrial Robot – Designation of Geometrical Axes and Movements* August, 1983.
AFNOR E 61-102 *Industrial Robot – Mechanical Structure* November, 1983.
AFNOR E 61-103 *Industrial Robot – Performance-Definitions* November, 1983.
AKR *The AKR Manual.*
Ambler, A.P.; Popplestone, R.J.; Kompf, K.G. An experiment in the off-line programming of robots. *Proceedings of 12th International Symposium on Industrial Robots* Paris, 1982.
Automatix *RAIL Software Reference Manual* February, 1982.
Bonner, S.; Shin, K.G. Comparative study of robot languages. *Computer* December, 1982.
Borrel, P.; Aldon, M.J.; Liegeois, A. CAD Robots LAM SITEF *CAD Journal Toulouse* October, 1981.
Borrel, P.; Bernard, F.; Liegeois, A.; Bourcier, D.; Dombre, E. The robotics function in CATIA. *State of Robotics in France* Volume 1, June, 1983.
Carrol, C.M. *The Great Chess Automaton* Dover, New York, 1975.
CAST *Industrial Robots and their Applications* Stage INSA, Lyon, 1982.
CETIM *Study of Manipulators and Industrial Robots* No. 4N-02-0, December, 1979.
Cincinnati Milacron *Operation Manual for the Cincinnati Milacron T3 Industrial Robot* (Version 3.0, Robot Control with Restructured Software) Publication No. 1-IR-79149, 1980.
Coiffet, P. *Modelling and Control* Robot Technology, Volume 1, Kogan Page, London, 1983.
Coiffet, P.; Chirouze, M. *An Introduction to Robot Technology* Kogan Page, London, 1983.
Engelberger, J.F. *Robotics in Practice* Kogan Page, London, 1982.
Friedman, A.D. *Logical Design of Digital Systems* Pitman, London, 1977.
Gini, G. *et al.* Program abstraction and error correction in intelligent robots. *Proceedings of 10th International Symposium on Industrial Robots* Milan, 1980.
GIXI *Reference Manual for ROL* May, 1982.
Gruver, W.A. *et al.* Evaluation of commercially available robot programming languages. *Proceedings of 13th International Symposium on Industrial Robots* Chicago, April, 1983.
Hasegawa, T. A new approach to teaching object description for a manipulator environment. *Proceedings of 12th International Symposium on Industrial Robots* Paris, 1982.

IBM *AML Concepts and User's Guide* Report SC34-0411-0, September, 1981.

IRIA International Seminar: *Languages and Methods for Programming Industrial Robots* Rocquencourt, June, 1979.

Khatib, O. Dynamic control of manipulators in operational space. *Sixth IFTOMM Congress on Theory of Machines and Mechanisms* New Delhi, December, 1983.

Komatsu *Robot Language PLAW* June, 1982.

Kretch, S.J. Advanced off-line programming for robots. *McDonnell Douglas Automation Company 2nd European Conference on Automated Manufacturing* Birmingham, May, 1983.

Laurgeau, C. *Industrial Programmable Controllers* SCM Editions, Paris, 1979.

Laurgeau, C.; Michel, G.; Espiau, B. *Programmable Industrial Controllers* Dunod, Paris, 1979.

Lewin, M.H. *Logic Design and Computer Organization* Addison Wesley, Reading, Massachusetts, 1983.

Lewis, A.O. *Of Men and Machines* Dutton, New York, 1963.

Lhote, F.; Kauffmann, J.; André, P.; Taillard, J. *Robot Components and Systems* Robot Technology, Volume 4, Kogan Page, London, 1984.

Libly, C. An approach to vision controlled arc welding. *Conference on CAD/CAM Technology in Mechanical Engineering.*

Llibre, M. *et al.* Data compression methods for the recording of industrial robot trajectories. *Proceedings of 12th International Symposium on Industrial Robots* June, 1982.

Mazer, E. LM-GEO geometric programming of assembly robots. *Imag Research Report* nb 296, March, 1982.

Mazer, E. An algorithm for computing the relative position between two objects from symbolical specifications. *Imag Research Report* nb 297, March, 1982.

McDonnell Douglas *Manufacturing Control Language User's Manual* January, 1982.

Miribel, J.F.; Mazer, E. *LM Reference Manual* Imag, October, 1982.

Mujtaba, S.; Goldman, R. *AL Users' Manual Report* STAN-CS-81-899, Stanford University, December, 1981.

Oldroyd, A. MCL: *An Approach to Robotic Manufacturing* SHARE 56, March, 1981.

Paul, R. *Robot Manipulators: Mathematics, Programming and Control* MIT Press, Cambridge, Massachusetts, 1981.

Prajoux, R.; Farreny, H.; Ghallab, M. *Decision and Intelligence* Robot Technology, Volume 6, Kogan Page, London (in preparation).

Queromes, J.G. Computer-aided design and robotics. *Proceedings of 12th International Symposium on Industrial Robots* Paris, June, 1982.

Salmon, M. SIGLA – The Olivetti Sigma Robot Programming Language. *Proceedings of 8th International Symposium on Industrial Robots* Stuttgart, May, 1978.

Sankyo *Robot Language SERF* 1981.

Sata, T.; Kimura, F.; Amano, A. Robot simulation system as a task programming tool. *Proceedings of 11th International Symposium on Industrial Robots* Tokyo, October, 1981.

Sigma Programming Handbook Olivetti, September, 1977.

Sjolund, P.; Donath, M. Robot task planning: programming using interactive computer graphics. *Proceedings of 13th International Symposium on Industrial Robots* Chicago, April, 1983.

Skoda, F. *LPR: Programming Languages for Robots* RNUR-DTAA, December, 1981.

Souza, Ch.O.; Zuhlke, D.; Blume, Ch. Aspects to achieve standardized programming interfaces for industrial robots. *Proceedings of 13th International Symposium on Industrial Robots* Chicago, April, 1983.

Taylor, R.H.; Summers, P.D.; Mayer, J.M. AML: A manufacturing language. *The International Journal of Robotics Research* 1982, **1** (3). IBM-T. J. Watson Research Center.

Unimation *Users' Guide to VAL, Version 12* June, 1980.

Unimation *VAL Univision Supplement, Version 13 (VSN)* July, 1981.

Unimation *VAL II Functional Specifications* January, 1982.

Villiers, P. The CAD/CAM link. *Industrial Engineering* 1982, **14** (4).

Winsh *et al.* Robex – An off-line programming language for controlling assembly robots. *Proceedings of 11th International Symposium on Industrial Robots* Tokyo, 1981.

Young, Y.F.; Bonney, M.C.; Knight, J.A. *Design and Simulation of Sheet Metal-working Flexible Manufacturing Systems* University of Nottingham, September, 1983.

Index